THE SCIENCE OF
SHERLOCK

THE SCIENCE OF
SHERLOCK

THE FORENSIC FACTS
BEHIND THE FICTION

MARK BRAKE

AUTHOR OF *THE SCIENCE OF DOCTOR WHO*

Skyhorse Publishing

Skyhorse Publishing books may be purchased in bulk at special discounts for sales promotion, corporate gifts, fund-raising, or educational purposes. Special editions can also be created to specifications. For details, contact the Special Sales Department, Skyhorse Publishing, 307 West 36th Street, 11th Floor, New York, NY 10018 or info@skyhorsepublishing.com.

Skyhorse® and Skyhorse Publishing® are registered trademarks of Skyhorse Publishing, Inc.®, a Delaware corporation.

Visit our website at www.skyhorsepublishing.com.

10 9 8 7 6 5 4 3 2 1

Library of Congress Cataloging-in-Publication Data is available on file.

Cover design by David Ter-Avanesyan
Cover image by gettyimages

Print ISBN: 978-1-5107-7057-7
Ebook ISBN: 978-1-5107-7058-4

Printed in the United States of America

This book is dedicated to my very own Irene Adler

CONTENTS

INTRODUCTION

AN ADVENTURE WITH THE DIAMOND OF DEDUCTION

. . . in which we meet a forensic tool for Sherlockian analysis.

"Any attempt at recovering the bodies was absolutely hopeless, and there, deep down in that dreadful caldron of swirling water and seething foam, will lie for all time the most dangerous criminal and the foremost champion of the law of their generation."

—Sir Arthur Conan Doyle, *The Adventure of the Final Problem* (1893)

The "Homicide" of Sherlock Holmes

Sherlock Holmes was dead. Or so it seemed. Murdered by the very man who made him—Scottish author Sir Arthur Conan Doyle. The murder weapon? Gravity. The crime scene? The Reichenbach Falls in Switzerland. Not even the world's greatest-ever consulting detective, "the last and highest court of appeal in detection," could survive a gravity-assisted plunge over a cliff . . . or could he?

It's December 1893, and Conan Doyle has decided to do the dirty work from his home in London, where he wrote most of his famous stories of not just the first fictional detective but also the best known. As narrator Dr. John Watson says in Conan Doyle's story *The Final Problem,* "It is with a heavy heart that I take up my pen to write these the last words in which I shall ever record the singular gifts by which my friend Mr. Sherlock Holmes was distinguished." *The Final Problem* appeared in *The Strand Magazine.* The public outcry at Holmes's death was unlike anything previously seen for a fictional character. And *The Strand* scarcely survived the resulting rush of subscription cancellations.

In private, there was merely the emotion of a cold-hearted assassin. "Killed Holmes," wrote Conan Doyle delightedly in his diary. It's easy to imagine Sir Arthur Ignatius Conan Doyle KStJ DL, Scottish-born writer of Irish descent, physician—and looking much like that other famous fictional detective, Hercule Poirot; glistening hair slicked back, twirling his bounteous mustache—as he relished every moment of Sherlock's murder. Much later, Conan Doyle was to confess, "I have had such an overdose of him that I feel towards him as I do towards paté de foie gras, of which I once ate too much, so that the name of it gives me a sickly feeling to this day."

The Phenomenon of Fandom

Sherlock was dead. A consulting detective no more. He'd ceased to be. Bereft of life, he rested in peace. His fictional metabolic processes now history, Sherlock had shuffled off his "mortal" coil and joined the crime scene invisible. At least, that was Conan Doyle's determination. But he hadn't figured on one important factor. Through the fiction of his Sherlock tales and their avid readers, Conan Doyle had helped create the modern phenomenon of fandom. Prior to Conan Doyle's creation, there had never been a pop culture character more famous than its creator. Sherlock was also the first fictional figure who inspired role-playing among his fans. Organized Sherlockian fandom began officially in 1934, with the establishment of the Sherlock enthusiasts' organization, the Baker Street Irregulars. And today, fan fantasy is a huge part of popular culture, from the Whovians of *Doctor Who*, to the Trekkers or Trekkies of *Star Trek*, and the Potterites of *Harry Potter*. These modern examples have their roots in the media fandom of the 1960s, which first coagulated around cult TV shows such as *The Man from U.N.C.L.E* and *Star Trek*. But Sherlock came first.

Sherlock's popularity and fame, and the force field that was his character on the page, were such that many believed him to be not a fictional fancy but a real man. Conan Doyle once confessed, rather uncharitably, "I get letters addressed to him. I get letters asking for his autograph. And I get letters addressed to his other stupid friend Watson." And as Roger Johnson, Officer of the Sherlock Holmes Society of London, said, "[Sherlock] was a man who was believable, even if he was unusual. And he lived in the world, in the city, that people knew. You might go out into Baker Street and see a tall thin man and you could believe that this was the real Sherlock Holmes." There was no way people would accept the great detective's death. Like all superheroes, Sherlock could never die.

The "murder" of Sherlock marked a black day indeed. Grief-stricken young men moped throughout London wearing black mourning crêpes on their hats, or around their arms, for the month following Sherlock's demise. In excess of twenty thousand *Strand* readers cancelled their subscriptions, outraged by Sherlock's untimely ruin. One typical letter to *The Strand* from a wrathful reader addressed Conan Doyle as a "brute!" In America, "Let's Keep Holmes Alive" clubs were created. The staff at *The Strand* came to refer to Sherlock's death as "the dreadful event." Yet, Conan Doyle stuck to his guns in this battle of wills. It was "justifiable homicide," according to Conan Doyle, though one suspects the justice meted out was his own, rather than that of Professor James Moriarty.

"Sherlock"

The "dreadful event" now sounds like a pretty typical day on the Internet in the twenty-first century. Indeed, January 2012 saw a contemporary rerun of the "death" of Sherlock Holmes. The BBC's flagship crime television series *Sherlock*, based on Conan Doyle's original detective stories, and starring Benedict Cumberbatch as Sherlock Holmes and Martin Freeman as Dr. John Watson, aired an episode called *The Reichenbach Fall* as the third and final episode of the second season. The episode attracted almost ten million viewers on its initial broadcast alone and became the second most-watched television program of 2012. The episode's cliffhanger resulted in a swarm of speculation (on forums, social networking sites, and in newspaper articles) about its resolution.

But today's commonplace subculture of fandom was nonexistent in the late nineteenth century. Conan Doyle had good reason to be stunned by the visceral reaction of his readership. "Fans" did not yet exist. Readers were supposed to take the latest tale on the chin and move on to another narrative. But, with Sherlock Holmes, readers began to take their culture to heart. They trusted their favorite characters and tales to hold to high expectations. They divined a kind of reciprocal relationship with the works they cherished.

The Creature of Conan Doyle

Conan Doyle and his readership helped forge the modern practice of fandom. Earlier in the nineteenth century, 1818 to be precise, Mary Shelley had written *Frankenstein,* the haunting tale of Dr. Victor Frankenstein's creation of a creature beyond his control. Dr. Conan Doyle created Sherlock in 1887's short novel, *A Study in Scarlet.* Sherlock was popular from the get-go, so much so that Conan Doyle began to regret having created his creature. In Doyle's words, "I've written a good deal more about him than I ever intended to do, but my hand has been rather forced by kind friends who continually wanted to know more. And so it is that this monstrous growth has come out of what was really a comparatively small seed."

The Sherlock tales haunted him. They cast a dark shadow over his other fictional work, work that Conan Doyle thought more worthy. As English novelist Anthony Horowitz puts it:

Conan Doyle became fed up with Sherlock Holmes, his greatest creation. Why? Because he thought he was a better writer. He was also fascinated by things like spiritualism, by politics, by travel, by the world. And he thought of Sherlock Holmes as sort of beneath him; an entertainment. I think it is interesting, the whole idea of the writer who finds himself hidden by his own creation, who finds himself smaller than his creation. The same thing, in a way, happened to Ian Fleming, who got rid of James Bond not once, but twice – [in] *From Russia with Love*, he's poisoned and is meant to die, and at the end of *You Only Live Twice*, he's got amnesia, he's vanished as a Japanese fisherman. But all the time these characters come back, they won't go away.

French psychoanalyst Pierre Bayard makes a different point:

Some characters, psychically very strong ones, escape from the creations and arrive in our world. And if we admit that things circulate between the real world and the world of fiction, then we may also ask ourselves whether we ourselves are not fictional characters. This uncertainty is what Freud calls the uncanny, played on by literature and mythical characters to the point that we can no longer be sure if they exist or not.

Sherlock "fans" queued up at newsstands whenever an issue of *The Strand* was to include a new Sherlock story. And it was solely down to Sherlock Holmes that one cultural historian wrote that Dr. Conan Doyle was as famous as Queen Victoria. Who were these avid Sherlock fans? Who were the shock troops of these early days of popular culture fandom? They were drawn from the emergent middle-class, the very social group whose culture would be belittled by snobbish critics as populist for many years to come. (The critics had been snobbish to Charles Dickens, too. Dickens's huge popularity during his own lifetime was followed by a reputational dip in the decades after his death, a decline attributed to the negative way that many literary critics viewed his huge achievements. George Henry Lewes, editor of the influential magazine *The Fortnightly Review*, noted the contrast between Dickens's "immense popularity" with what he called "critical contempt." "There probably never was a writer of so vast a popularity," he wrote, "whose genius was [so] little appreciated by the critics.")

These early Sherlockians were looked down upon. They were priceu ut of fashionable music concerts. And they had to hang around for the bargain-basement editions of popular novels. They were mostly drawn from the lower middle classes of the growing cities—nonintellectual, non-private school, but industrious, rising, emergent. *The Strand* had them in its sights. They published thrilling and big-idea genre stories, mysteries, and science fiction, from authors such as Jules Verne, H. G. Wells, and Conan Doyle himself. The clamor for Sherlock stories seemed insatiable. And one important feature of serialized storytelling that happened to have populist potential was the opportunity it gave to readers and friends to exert influence over a story's outcome. Anyone so inclined could exert pressure on the author before the next installment was published. *The Strand* paid Conan Doyle handsomely for every Sherlock tale he told. But he hadn't intended a career creating and decoding fictional crimes. Conan Doyle was merely meant to be making money to fund his "real" art—writing political novels full of what he felt were weighty ideas.

A key ingredient in Sherlock's success was the serial nature of his adventures; Conan Doyle's was an early sequential art. This ingredient explains why the first two tales, *A Study in Scarlet* and *The Sign of Four*, were merely modest successes in their novel format. But Sherlock's popularity only truly soared with the move to the monthly short-story adventures in *The Strand* from 1891 on. It was arguably one of the most effective innovations in literary publishing history.

In his *Memories and Adventures*, Doyle looked back on the marriage of form and content that resulted in the unprecedented popularity of Sherlock:

> A single character running through a series, if it only engaged the attention of the reader, would bind that reader to that particular magazine. On the other hand, it had long seemed to me that the ordinary serial might be an impediment rather than a help to a magazine, since, sooner or later, one missed one number and afterwards it had lost all interest. Clearly the ideal compromise was a character which carried through, yet installments which were each complete in themselves, so that the purchaser was always sure that he could relish the whole contents of the magazine. I believe that I was the first to realize this and *The Strand Magazine* the first to put it into practice.

Doyle's reading public adored the format of the monthly episodes for good reason. Think about the tales' criminological "context." Sherlock fans understood there would be a new tale each month, which not just raised questions about the finality of each adventure, but also mirrored a Victorian obsession that crime too was intrinsically repetitious; during the 1890s, around 55 percent of prisoners were repeat offenders, a figure which rose to 75 percent early in the new century. Doyle tried his best, within his story formula, to avoid featuring repeat offenders, yet he created modern culture's paradigmatic and iconic repeat offender in the criminal mastermind Professor James Moriarty. Moriarty's relatively elusive presence in Conan Doyle's tales—he is alluded to in several stories but appears in person just once—has since been counterpoised by his appearances in so many of the Sherlock reimaginings and remixes.

Sherlock Holmes Lives On

Sherlock Holmes simply wouldn't die quietly. Conan Doyle had been only thirty-four when he had Moriarty send Sherlock reeling down the Reichenbach. But, eight years later in 1901, fan pressure had mushroomed so much that Conan Doyle was forced to write the now-famous *Hound of the Baskervilles*, a story which featured a Sherlock before his fall. Conan Doyle had written, "I heard of many who wept. I fear I was utterly callous myself and only glad to have the chance of opening out into new fields of imagination, for the temptation of high prices made it difficult to get one's thoughts away from Holmes."

A mere two years later, and Conan Doyle had resurrected Sherlock completely. There had been an uproar after the publication of *Hound of the Baskervilles*, and a tremendous amount of pressure to bring Sherlock back. The American magazine *Colliers* tried to seduce Conan Doyle with five thousand dollars per story, plus royalties. Then he was offered thirty thousand dollars for six stories, and sixty-five thousand dollars for thirteen. Impossible to refuse, Sherlock was fully resuscitated in 1903's tale *The Adventure of the Empty House*, and the Sherlockians learned that only Moriarty had died in the falls—Sherlock had faked his own death. To explain his hero's extended absence, Conan Doyle invented a myriad of adventures during this period known as the "Great Hiatus." Sherlock had changed identities. He'd met the Dalai Lama in Tibet. Explored Norway. Crossed Persia. Visited Mecca. Lived in Khartoum.

If anything, Sherlockians have become more fanatical since those early days. The BBC's *Sherlock* met with huge success. On some occasions, hundreds of fans would show up at London locations merely to watch an episode in the making. Praised for the quality of its writing, acting, and directing, *Sherlock* was nominated for a variety of media awards, which included Emmys, BAFTAs, and Golden Globes, winning awards across a range of categories. It was a commercial success, too. Produced at BBC Wales (just down the road from my home), *Sherlock*'s third series became the UK's most watched drama series in over a decade, and the production as a whole has been sold to over 180 territories.

Make no mistake. *Sherlock* is firmly based on Conan Doyle's creation. The show is, in a very real sense, fan fiction founded on Conan Doyle's Victorian-age work. A critic for *The Guardian* declared the series to be "brilliantly promising" and "indisputably Sherlock Holmes." Such success was put down to the writers being "enormously knowledgeable about Conan Doyle's work, and their reimagining incorporates big- and small-screen adaptations of Holmes." And a critic for *The Telegraph* said that "Cumberbatch is utterly credible as a man who lives entirely in his cerebellum with little regard for the world outside, mak[ing] *Sherlock* the perfect depiction of Holmes for our times."

Sherlock remains with us still. As Anthony Horowitz puts it, "Conan Doyle invented the modern detective story. All modern detective stories begin with Sherlock Holmes. Just the very idea of the three-act book, starting with a murder, the investigation of a solution, self-contained in that way, nobody had done it before Doyle."

Little wonder Sherlock has gone global. Fan fiction in China. Sherlock manga in Japan. Tribute pop songs in Korea. It's a continuation of the fan adoration over a fantastical detective who has lasted not far short of one and a half centuries, and through many adaptations. In 2012, *Guinness World Records* awarded Sherlock Holmes the title of most portrayed literary human character in film and television. Holmes has been played by over seventy-five actors, including Sir Christopher Lee, Charlton Heston, Sir Ian McKellen, Peter Cushing, Sir Michael Caine, Peter O'Toole, Christopher Plummer, Peter Cook, Sir Roger Moore, John Cleese, Benedict Cumberbatch, and Robert Downey Jr. Award adjudicator Claire Burgess said, "Sherlock Holmes is a literary institution.

This *Guinness World Records* title reflects his enduring appeal and demonstrates that his detective talents are as compelling today as they were 125 years ago."

Sherlock as Part of a Remix Culture

After the Bible, the Sherlock stories are the most widely circulated and translated books in the world. This enables us to reimagine Conan Doyle's creation as a retro example of *remix culture*. Remix culture, also known as read-write culture, is a term that describes a society which permits and endorses derivative art by combining or editing existing materials to make new creative works. With almost three hundred films to date, more than one thousand television episodes, and just as many imitations, let alone the video games and comic books, Sherlock has been remixed more frequently than Dracula, Frankenstein, Napoleon, or Jesus Christ. Across the ages and continents, Sherlock is the literary remix par excellence. (Figure 1 on page 11 shows data from the movies alone.)

Consider the evidence. The line "Elementary, my dear Watson" appears not once in the canon of Conan Doyle's sixty Sherlock tales. Sherlock's main rivals in the cultural recognition stakes—Count Dracula and Frankenstein's creature—are mostly imagined in terms of movie adaptations rather than their literary originals. Sherlock is a remix amalgam, a blend way beyond the reach of any single adaptation or representation, including Doyle's original. Let's think about that. In the popular imagination, Mary Shelley's monster still has the face of Boris Karloff from the 1931 movie, *Frankenstein*. And Dracula is almost always hugely derived from Bela Lugosi's iconic portrayal of the Count in the film *Dracula* of the same year. But Sherlock's popular mashed-up and remixed persona is part from the pen of Conan Doyle, part the illustrations of Sidney Paget, part the theatrical adaptation of William Gillette, part the movie portrayal of Basil Rathbone, and part the televisual renderings of Jeremy Brett and Benedict Cumberbatch. Each new generation

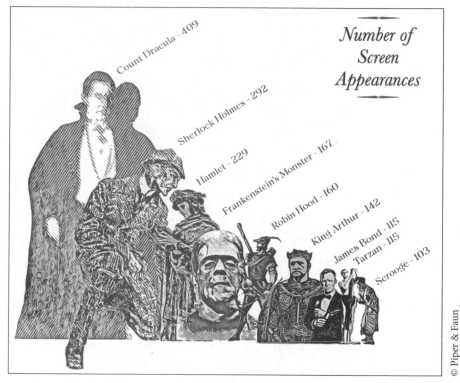

Number of Screen Appearances

Count Dracula - 409
Sherlock Holmes - 292
Hamlet - 229
Frankenstein's Monster - 167
Robin Hood - 160
King Arthur - 142
James Bond - 115
Tarzan - 115
Scrooge - 103

© Piper & Faun

Figure 1. Sherlock's silver screen appearances.

may have its preferred image of Sherlock, but no one preference subsumes the rest; no version is definitive.

Another key influence on the Sherlock remix potential is the serial nature of the adventures and the intrinsically repetitious format of Doyle's tales. This repetitious structure allowed narrative freedoms to liberate Sherlock from the tethers of the Thames and the bonds of Baker Street and set him free into the wider world. Compare this with the likes of Frankenstein's creature and Dracula, whose origin stories are somewhat set in the narrative stone laid out for them by Mary Shelley and Bram Stoker.

The diverse medley of Sherlock's sixty stories fostered the creative remixes that followed. Doyle's familiar formula became enabling rather than limiting. Sherlock and Watson could be transplanted into alien scenarios, other places, other fictional universes. And future writers of Sherlock tales could take the general timeline of a Doyle text (a beginning on Baker Street, a ritual display and timely reminder of our detective's deductive powers, the client coming to consult on a case, etc.) and change the details to suit their new narratives. Conan Doyle's creative combination of formal familiarity with kaleidoscopic content rendered Sherlock particularly pliant to the early signs of fandom bubbling up in the 1890s. Little wonder that an early manifestation of this fandom, during the Great Hiatus after Sherlock's "death" in 1893, was a competition in *The Strand*'s familial publication, *Tit-Bits* (yes the magazine *is* seriously called that, as the British really are quite odd), inviting its readers to pen their own Sherlockian tales.

Such remixing writing contests of Sherlock adventures also helped advertise *The Strand*. Sherlock was, after all, the strongest brand they possessed. And boy did that brand evolve. From being the commodity for sale in the 1890s, Sherlock developed during the twentieth century into the means of the advertisement itself. The examples of companies and commodities are too numerous to mention but suffice it to borrow from Amanda J. Field's *Sherlock Holmes in Advertising*, which included New Golden Glow Beer, Teachers' whisky, the Yellow Pages, Canon typewriters, Kellogg's Crunchy Nut Cornflakes, and Kodak.

Whereas Doyle's rendering of Sherlock arguably codifies certain forms of Englishness, masculinity, and the scientific method, in the face of the commodity fetishism of capitalism, he is also an amazingly remixable character. In the words of Amanda J. Field, Sherlock is "a floating signifier that can be applied at will to different advertising campaigns in different historical situations." A *Mad Men* dream.

Sherlock and Science Fiction

The remix adaptability of Sherlock may center around his status as a "floating signifier." Sure, Doyle created Sherlock as a decisive figure of detective fiction. But Sherlock's mutability meant that he could be remixed into other genres and modes. Many believe that Sherlock is just as much a mainstay of science fiction as he is of detective fiction. This argument cuts two ways; first, that Conan Doyle's texts can be reappraised as science fiction, and second, that Sherlock can be creatively recast in a sci-fi mold.

For the first argument, it's pointed out that a Sherlock tale like *The Adventure of the Creeping Man* can be read as sci-fi in the way that it extrapolates fantastic outcomes from prevailing scientific theories (in the case of *Creeping Man*, the theories concerned relate to degeneration anxieties). For the second argument, it's noted how often Sherlock pops up in texts which are identifiably science fiction. For example, in the late–twentieth century there was a recurring theme of a cryogenically de-animated Sherlock, suspended until he is resuscitated and reanimated in the future, whether that future is his or ours. Such is the plot of movies like 1987's *The Return of Sherlock Holmes*, 1993's *Sherlock Holmes Returns*, and especially the animated series *Sherlock Holmes in the 22nd Century* (which ran from 1999 until 2001), where the "born again" Sherlock is assigned an automaton companion named Watson. (The irony here is that Watson had described Sherlock as an automaton when Sherlock failed to recognize Mary Morstan as an attractive woman in *The Sign of Four*: "[Y]ou really are an automaton—a calculating machine . . . There is something positively inhuman in you at times.") Furthermore, AI representations of Sherlock's universe play a vital part in episodes of *Star Trek: The Next Generation* (1988's *Elementary, my Dear Data* and 1993's *Ship in a Bottle*), which look at human perception and the extent to which literary worlds like Doyle's are virtual realities in themselves.

Sherlock and Doctor Who

An interesting sci-fi remix is that which riffs on the parallels between Sherlock and the Doctor in the BBC's science-fiction television program, *Doctor Who*, which itself has run through various remixes and regenerations since 1963. In *What is Doctor Who?*, a blog post written by Adam Roberts, British science fiction novelist and professor of nineteenth century literature at Royal Holloway, University of London, Roberts's argument is that as he initially appeared, and through most of his incarnations, the Doctor is a *gentleman*. Roberts explains:

> What *is* the Doctor? He is a man of breeding and wealth (the two things don't always go together, but in this case they do), factors that enable him to evade the responsibilities of work that bear down upon the rest of us . . . two other features of the Doctor's personality that are not only gentlemanly but more specifically Victorian- or Edwardian-gentlemanly: he is *eccentric*, and he is *not a snob*. Eccentricity is a marker of class in the practical sense that a gentleman can get away with acting oddly and indulging his personal crotchets in a way that would lead to a working man (or woman) losing their jobs, or being otherwise socially sanctioned.

Roberts then compares the Doctor to Sherlock whose eccentricity is "absolutely part of his gentility." There are plenty of remix references to Sherlock in the *Doctor Who* canon. The heroes meet in stories such as 1994's *All-Consuming Fire* by Andy Lane. In 1977 there was a *Doctor Who* episode entitled "The Talons of Weng-Chiang." It was set in the society of London's theaters at the *fin de siècle* and has the Doctor popping on an imprecise deerstalker and cape on the hunt for a killer and, no coincidence here, a giant rat, which echoes Watson's allusion to the giant rat of Sumatra, "a story for which the world is not yet prepared" (*The Adventure of the Sussex Vampire*). What's more, the Sherlockian Doctor of Talons was played by Tom Baker, the Fourth Doctor, who would later portray Sherlock in the BBC's *Hound of the Baskervilles* in

1982. The BBC's *Sherlock* is also produced by two writers with a major role in *Doctor Who*'s modern revival, Steven Moffat and Mark Gatiss.

This similarity between Sherlock and the Doctor should not surprise us. After all, *Doctor Who* was largely based on the 1895 H. G. Wells novel *The Time Machine*. Wells's story is incredibly important to *Doctor Who*. The Doctor uses a space-time vessel, in the guise of the TARDIS, to time-travel at leisure, just like the Time Traveler in Wells's original *Time Machine*. But it goes deeper than that. The character of the original Doctor as a Victorian, or Edwardian, gentleman was based on Wells's Time Traveler, and the Time Traveler has great similarities with Sherlock.

The Time Traveler and Sherlock were created a mere seven years apart. Writers like Wells and Doyle, educated in science at the Normal School of Science in London and Edinburgh Medical School respectively, sought to explore and unmask scientific truth in their tales, to solve profound mysteries in various situations and settings. Wells's way of doing so was to practically invent modern science fiction in tales such as *The Time Machine*. Doyle's way was to create detective fiction, conjuring up his iconic detective in Sherlock.

Wells's Time Traveler, and the Doctor, are detectives in that they concern themselves with the profound mysteries of the future and ultimately that of humankind. Meanwhile, Sherlock is London's more local savior, its Newton of crime. Sherlock busies himself with the mysteries of less significant "trifles" in the world's first industrialized city. A different kind of alien landscape. All three fictional characters are detectives of science, even if their methods of reasoning, imaginative thinking, and hypothesis-making differ. The Time Traveler and the Doctor journey through the darkness of space to decipher the mystery of humanity's fate against the relentless tide of time. Sherlock journeys through the dark streets of London to secure the fate of his city against the rising tide of crime.

This close remix relationship between Sherlock and *Doctor Who* has another, still deeper, resonance. British culture, rooted in a Celtic tradition, requires that its heroes be resurrected, from the pledged reappearance of King Arthur to the regenerations of the Doctor. As Conan Doyle wrote, "Why should we fear a death which we know for certain is the doorway to unutterable happiness? Why should we fear our dear one's death, if we can be so near to them afterwards?" In trying to murder Sherlock in 1893, the fact of his resurrection in *The Adventure of the Empty House*, meant that Conan Doyle simply served to cement Sherlock's place as a truly mythic figure in British culture.

We live in a remix universe where Sherlock has left his mark everywhere. Cocreator of the BBC's *Sherlock*, Mark Gatiss, put it this way:

> [Sherlock] has a Victorian superpower, which still works in the modern day. I think essentially it comes down to the fact that he is the smartest man in the room. He can make the connections nobody else can. And *that* is timeless. It must speak of a kind of need we have to be saved, I think, or to believe there is something slightly higher than us which is going to come and get us out of this terrible mess we're all in!

British writer and broadcaster Matthew Sweet puts it another way:

> [Sherlock]'s the most depicted fictional character, certainly the most *filmed* fictional character. I think it's because of the strange sort of glamour that he possesses. It's a rather uneasy sort of glamour because he is a glacial and frightening character. But he's rather like a character from gothic fiction. He's rather like a vampire . . . There's [also] something Christ-like about [Sherlock]. He's the man who dies and rises again. He

falls quite a long way, but he comes back, and I think that's one of the reasons why he inspires that kind of following.

Dark glamour and a Christ-like character are two aspects of Sherlock that have left their permanent mark on popular culture. Another aspect is that suggested by British intellectual Stephen Fry: "Holmes meant the world to me when I was young. And I think what most . . . Sherlockians love is that mixture of detail and authority, that sense of wisdom; the world being a solvable thing. We all want, we'll *search* for, a teacher, a master, someone who can be that figure to us."

The word remix originally referred to music. It emerged in the late twentieth century during the heyday of hip hop, which was the first popular music form to integrate sampling from existing recordings. An early example is the Sugar Hill Gang's sampling of the bass riff from Chic's recording "Good Times" for their huge 1979 hit "Rapper's Delight." Since then, the Chic bass line has been sampled dozens of times. You can see a literary similarity with Sherlock.

But remixing, or whatever we might call it, didn't start with hip hop. Earlier musicians remixed too, through copying and homage. In the early 1970s, British rock band Led Zeppelin became hugely famous for innovating a new kind of incredibly loud electric blues and, within just a few years, became the biggest band on the planet. But Led Zeppelin also "remixed." Much of their source material was drawn from traditional black blues musicians many years before. Zeppelin simply did what all artists do: Copy from others, transform those ideas, and combine them with other ideas to create a new synthesis.

It soon became clear that artists had been sampling for centuries. This realization was contrary to the traditional idea that creative art was somehow divinely channeled from God, creating works of singular genius without any cultural or societal influence. Consider one of Pablo Picasso's most famous works, *Les Demoiselles d'Avignon*. Painted in 1907, while Conan Doyle was still in full flow with his creation of Sherlock stories, *Les Demoiselles d'Avignon* portrays five nude female prostitutes in a Barcelona brothel. Each prostitute is shown in an unsettling and confrontational manner, while none is conventionally feminine. In fact, the prostitutes are slightly menacing and rendered with angular and disjointed body shapes.

Les Demoiselles d'Avignon is still thought to be seminal in the early development of both cubism and modern art. It was considered the gold standard for creativity because it was thought to be unprecedented; nobody had seen anything like it before. But when you dig a little deeper into Picasso's famous painting, the signs of remix are clear. The figure on the left shows facial features and dress of Egyptian or southern Asian style. The two adjacent figures are depicted in the Iberian style of Picasso's native Spain, and the two on the right are portrayed with African mask-like features. Indeed, according to Picasso, the ethnic primitivism evoked in these masks inspired him to liberate "an utterly original artistic style of compelling, even savage force." The Picasso example shows that beneath the myths of creativity lies a more profound reality of remix.

Consider Conan Doyle's creation of Sherlock Holmes. To what extent was the original Sherlock a remix? He is a paper creation born out of a writer's brilliant imagination, and he's certainly been the subject of many remixes since. What makes Sherlock timeless, and the reason he's been the subject of so many remixes, is that writers came to realize that Sherlock didn't have to stay in Victorian London. He could change countries, places, periods. He's been set against Jack the Ripper, Dracula, Frankenstein. Adapted into hundreds of films and parodies, he has appeared in every media imaginable, and each time he's different yet unchanged.

What were the derivatives and existing materials from which Conan Doyle conjured his new creative work? This question will be answered on page 49, where we look at the four personality factors which fed into the creation of Sherlock Holmes: Conan Doyle himself, Edgar Allan Poe, Dr. Joseph Bell, and Sir Isaac Newton.

Remix: From Holmes to *Sherlock*

In the last decade or so, Sherlock Holmes has been remixed in two very popular forms. The first comes from British film director Guy Ritchie. His fantasy Victoriana films, *Sherlock Holmes* and its sequel *Sherlock Holmes: Game of Shadows*, are Oscar-nominated and star Robert Downey Jr. as Sherlock and Jude Law as Watson. The second, of course, is the BBC's *Sherlock*.

There is even resonance between these two remixes. Season three of *Sherlock* showed the continuing and amazing adaptability of Conan Doyle's fictional character by integrating some of the style of the two Guy Ritchie films, along with various elements of the hugely pliable Sherlock canon. These two remix adaptations showed how happy we all are at the prospect of living with different iterations of Sherlock, and even happy when they adapt and reference one another (witness Benedict Cumberbatch's action-hero window smash in "The Empty Hearse," followed by a nonchalant hair tousle and a Hollywood kiss with Molly Hooper).

The BBC series is also canny about its Victorian origins, especially when it comes to remixing the costumes of the nineteenth century in a modern setting. Upon Sherlock's return to London, we viewers are greeted with a panoramic and Romantic scene of our hero, surveying his city from on high, clad in his now iconic greatcoat; a scene which recalls the powerful image of the nineteenth century gentleman explorer. This idea of Sherlock surveying London from on high is one we shall return to on page 33, where we see the evolution of the city through the eyes of Sherlock himself.

Sherlock in the Machine Age

The original world of Sherlock Holmes is the clanging new workshop of the world that was Victorian Britain. "Were we required," wrote Scottish historian Thomas Carlyle in 1829, "to characterize this age of ours by any single epithet, we should call it the Mechanical Age." And, as the machinery began to mesh, science encroached upon all aspects of life, meeting every challenge with a new invention. The steam engine drove locomotives along their metal tracks. The first steamships crossed the great Atlantic. Transport magnates built bridges and roads. Telegraphs ticked intel from station to station. Cotton works glowed by gas. And a clamorous arc of iron foundries and coalmines powered this Industrial Revolution.

Sherlock Holmes was spun on the crackling loom of this machine age. Mary Shelley had Victor Frankenstein make his "monster" with the new science in his "workshop of filthy creation." The new philosophers of science seduced Victor: "They ascend into the heavens: they have discovered how the blood circulates, and the nature of the air that we breathe. They have acquired new and almost unlimited powers; they can command the thunders of heaven, mimic the earthquake, and even mock the invisible world with its shadows." Those new philosophers of science seduced Conan Doyle, too. His "workshop of filthy creation" was the popular fiction of big-idea genre mysteries, and his creature was Sherlock, an uber-rational character too scientific for some tastes, with a cold-blooded passion for definite and exact knowledge.

Sherlock became a major figure of the machine age. Voltaire once wrote, "If God did not exist, it would be necessary to invent Him." The same is true of Sherlock, especially in a Victorian age where a scientific vision of material progress had been finally realized. This is why Sherlock became so successful. Conan Doyle made flesh something that had not been previously articulated. Sherlock personified the spirit of the times. The desire to believe in science's omnipotence in an era that had been totally transformed by the captivating but troubling Industrial Revolution.

Charles Dickens had been able to describe London as a city that was acquisitive and hungry but where there was still some goodness to be found. In contrast, Conan Doyle drew a dark diagram of London, a city where crime pervaded the streets and many people lived in its shadows. What was needed now was a hero who understood the dark side of human life as well as any criminal. What Gotham City is to Batman, Victorian London is to Sherlock; is this not the same city as stalked that joker Jack the Ripper? It's down these gothic streets Sherlock prowls at night, disguised. It's his city's diabolical crimes and conspiracies that demand all the power of his insight and deduction—to meet atrocious power that often verges on the inhuman. And Sherlock's startling analytical and deductive virtuosity came with the promise of the eventual triumph of reason and the end of darkness.

Sherlock Holmes is the definitive man of science, an innovator of forensic methods. He is so much at the bleeding-edge of detection that he has written many monographs on crime-solving techniques. Conan Doyle often has Sherlock employing methods years before they were used by official police forces in both Britain and America. The result was sixty stories in which logic, deduction, and science dominate detection methods.

With Sherlock also begins that convention of so many fictional crime solvers: the brilliant-but-broken formula. His whimsies of manner and habit, from playing the violin badly to keeping his cigars in the coal scuttle and his tobacco in the toe-end of a Persian slipper, and . . . taking cocaine. As British intellectual Stephen Fry puts it:

What we love about Holmes is that he was flawed. And Conan Doyle seemed to go to great lengths to show that he *was* flawed; as a social animal, as it was in terms of his relations with women and society—in general quite awkward and difficult. And flawed in his drug addiction, the famous seven percent solution of cocaine, and so on. And [he was] occasionally impatient and irascible, difficult, subject to moods. It's quite clear from the early stories, in particular, that he's drawing a picture of someone's bipolar disorder, actually. Because he does leap from the slough of despond to a sort of merry prankster, and not sleeping—that sort of thing.

In crime fiction there are so many clones of Sherlock. Even outside the world of detection, Doyle appears to have started this fashionable idea that superintelligence comes at the price of some kind of social dysfunction. Sherlock is a genius, but a bit strange. (Just think of the legions of superintelligent weirdos that followed the example of Sherlock: Hercule Poirot, Lieutenant Columbo, Dr. Emmett Brown, Dr. Walter Bishop, Sheldon Cooper, Spock, Doctor Who, Dr. Gregory House, and various mad doctors of the movies, including the doctors Strangelove, Rotwang, and Lecter, to name but a few!)

We are living in Sherlock's world. So many creators of fictional universes have been inspired by Conan Doyle's example. Sherlock's genius is to take a dark disorder of ominous events and, by the brilliance of his scientific intelligence, flood it with light. It's all about illumination. After all, darkness can't drive out darkness; only light can do that. The idea of using fiery light to fight the darkness has been with us for centuries. Fire and light have illuminated our journey through the world. Light is a symbol for hope and enlightenment. Take it on your adventures and light is the spark of wisdom and knowledge, the flame of which is to be kept alight at all times. And, with Sherlock, there is a thin line between light and dark, between insight and insanity.

You can see why some say that all modern science fiction, mystery fiction, and genre fiction is a remix of Sherlock and his adventures. Sherlock's fingerprints are on every mystery novel, every detective character, and every conflict between the rational and the irrational, from *X-Files* to *Lost* to most police procedural dramas. It all comes back to Sherlock. In some way or another, Sherlock is a character who conjured something so potent and so powerful that his clones were remixed, and their iterations went running through the culture.

A huge amount of Sherlock's influence also comes down to his relationship with Watson. The dynamic duo of Sherlock and Watson, their relationship, the milieu they inhabit, the friendship between the two men, is the prime influence of the buddy film genre. The duos in these movies tend to have the same kind of interplay that Sherlock has with Watson. The pairing of two men, one distant and unknowable, the other much more like us, is hugely influential. You see the same kind of dynamic in movies such as *Lethal Weapon*, *Butch Cassidy and the Sundance Kid*, in the relationship between Captain Kirk and Mr. Spock in *Star Trek*, etc. But before all of them there was Sherlock and Watson. They're the prototype of a century's movie remix of the Conan Doyle kinship.

Indeed, the Sherlock-Watson relationship remains famous in several ways. And we can still find evidence of an astute Sherlock, and a rather comedically dumb Watson, in today's culture. In 2001, *The Guardian* newspaper carried reports on *LaughLab*, a twelve-month University of Hertfordshire project to find the world's funniest joke. *The Guardian* wrote that, in the first quarter of the project, "more than 100,000 people from seventy countries have visited the *LaughLab* website, submitted a total of 10,000 jokes, and rated them on a specially designed 'laughometer.'" At the time of *The Guardian*'s report, the leading joke (with 47,000 votes) featured the Sherlock-Watson relationship: Sherlock Holmes and Dr. Watson are going camping. They pitch their tent under the stars and fall asleep. In the middle of the night, Sherlock wakes Watson up: "Watson, look up at the stars, and tell me what you deduce."

Watson: "I see millions of stars and even if a few of those have planets, it's quite likely there are some planets like Earth, and if there are a few planets like Earth out there, there might also be life."

Sherlock: "Watson, you idiot, somebody's stolen our tent!"

The gag works on various levels. Even those with the slightest knowledge of Sherlock and Watson will at once recognize their archetypal characters: Sherlock the smart-ass, Watson the doughty companion. Others who know the Sherlock canon well will maybe grok the nod to Sherlock's own camping expedition in *The Hound of the Baskervilles* and his alleged ignorance of the workings of the Solar System in *A Study in Scarlet*.

One final and important contribution to the dynamic duo of Sherlock and Watson is the impact of the visual culture of Sidney Paget's illustrations. Paget's images were not only instrumental in cementing Sherlock's popularity, but they also almost completely focus on people and portraiture, especially of Sherlock. Paget's portrayal of Sherlock was based on his brother Walter. A portrait hugely more handsome than Conan Doyle's initial conception of Sherlock as having "a thin, razor-like face, with a great hawk's-bill of a nose, and two small eyes, set close together on either side of it."

Of the 201 Paget illustrations created for the *Adventures* and *Memoirs* story collections, a full 121 show Sherlock, 82 of which also include Watson. Sherlock is frequently pictured alone, whereas Watson gets this special treatment only twice. But Watson's authority as our reliable narrator is subtly underlined by Paget in that both story collections close with his image, not Sherlock's.

The Deduction Diamond

In this book, we shall look at the science of Conan Doyle's works. But, as you may have already gathered, we shall also look at the way in which science and society might have affected the reader, and indeed Conan Doyle himself. A fuller understanding of Conan Doyle's works is dependent on having knowledge about the circumstances in which they were produced. This might include information about the creator(s) who produced them, when and where they were actively writing, and what was happening in science and culture at the time, either locally or in the world at large.

When thinking of the works and many adaptations of Conan Doyle's Sherlock tales, the word "text" covers more than just the written word of Conan Doyle. A text can be an original Sherlock tale by Conan Doyle, sure, but also one of the many movie or television adaptations, such as the BBC's *Sherlock*, artistic representations of the tales, interviews with Conan Doyle himself, articles about Conan Doyle and his creation, and so on. Here's a proposal. When considering a Sherlock text, try using the Deduction Diamond (see Figure 2 below) to help you consider the science in the sixty Sherlock Holmes stories and associated texts. This analytical tool helps strike an easy balance between your personal views on Conan Doyle's work and the wider context of science and culture that aren't necessarily highlighted on its pages.

The Deduction Diamond is a method used for looking at fiction, as well as other aspects of art and history, which gives you a reliable and reusable formula for coming to sound conclusions about particular works. The four points of the Deduction Diamond represent particular tasks that, in combination, lead to satisfying and thorough interpretations about works of fiction, and especially

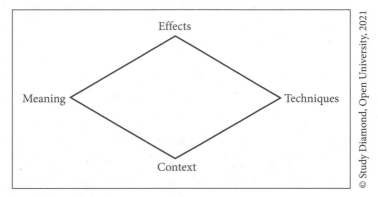

© Study Diamond, Open University, 2021

Figure 2. Deduction Diamond.

Conan Doyle's Sherlock tales. As you can see, the four points are labeled "effects," "techniques," "context," and "meaning," and all relate to one another.

Here's how the diamond is a four-step way of looking at any text. It allows you to:

- Think about the *effects* the text has, what it makes you feel or think about
- Consider the *techniques* Conan Doyle used to create the effects
- Think about the factual *context* within which Conan Doyle's text was made
- Consider the possible *meaning* of the text (what's it all about?)

The points of the Deduction Diamond can be considered in any order. Some people like to start with effects and move around the diamond in a clockwise direction, taking each point as they go. But the most crucial thing is to consider all four points of the diamond and reach a balanced and well-argued deduction.

The best introduction to the Deduction Diamond is to simply try it out. Read the passage below. It's taken from Conan Doyle's first story: 1887's short novel, *A Study in Scarlet*. Dr. Watson, our narrator, is writing about his newfound friend, Sherlock Holmes. At the time Watson made these remarks, he was unaware of Sherlock's work as a consulting detective:

His ignorance was as remarkable as his knowledge. Of contemporary literature, philosophy and politics he appeared to know next to nothing. Upon my quoting Thomas Carlyle, he inquired in the naivest way who he might be and what he had done. My surprise reached a climax, however, when I found incidentally that he was ignorant of the Copernican Theory and of the composition of the Solar System. That any civilized human being in this nineteenth century should not be aware that the Earth travelled round the Sun appeared to be to me such an extraordinary fact that I could hardly realize it . . . He said that he would acquire no knowledge which did not bear upon his object. Therefore all the knowledge which he possessed was such as would be useful to him. I enumerated in my own mind all the various points upon which he had shown me that he was exceptionally well-informed. I even took a pencil and jotted them down. I could not help smiling at the document when I had completed it. It ran in this way:

Sherlock Holmes—his limits.
1. Knowledge of Literature—Nil.
2. Philosophy—Nil.
3. Astronomy—Nil.
4. Politics—Feeble.
5. Botany—Variable. Well up in belladonna, opium, and poisons generally. Knows nothing of practical gardening.
6. Geology—Practical, but limited. Tells at a glance different soils from each other. After walks has shown me splashes upon his trousers, and told me by their color and consistence in what part of London he had received them.
7. Chemistry—Profound.
8. Anatomy—Accurate, but unsystematic.
9. Sensational Literature—Immense. He appears to know every detail of every horror perpetrated in the century.
10. Plays the violin well.
11. Is an expert singlestick player, boxer, and swordsman.
12. Has a good practical knowledge of British law.

The Deduction Diamond: Effects

Using the points of the Deduction Diamond, let's start first with the *effects*. What impression of Sherlock does Conan Doyle give us here? Now, bear in mind that Watson doesn't know Sherlock particularly well at this point. Nonetheless, Watson rather precisely portrays for us a kind of encounter with a superhero. Sherlock is a remarkable individual, with a profound knowledge in some areas (an expert boxer and swordsman with a complete knowledge of chemistry makes Sherlock sound like a pugilist Dr. Jekyll!), yet this superhero has his limits, seeds of normality, which make him feel more human.

The Deduction Diamond: Technique

What *technique* does Conan Doyle use to convey such effects? The main literary method used here is the tabulation, in bullet points, of Sherlock's abilities and skills, albeit subtitled "his limits." The use of the list is a kind of anatomy, an attempt to examine all parts of a subject, and reminds us that Watson is, of course, medically trained and has set out the facts with a clarity that helps his precise analysis.

The Deduction Diamond: Context

What about the factual *context* within which Conan Doyle's text was written? What was the state of science in 1887? Welcome to the machine age. Technological progress proceeded apace. In Austria, an astronomical compilation of the 8,000 solar and 5,200 lunar eclipses from 1200 BC until 2161 AD was published, stressing the predictive power of the scientific method. In Germany, the Petri dish was created, highlighting developments in experimental science. A Romanian chemist first synthesized amphetamine this year, enabling better understanding of the human central nervous system. In climate science, the largest snowflakes are recorded (reported during a snowstorm at Fort

Keogh, Montana, at 15 inches wide and 8 inches thick), enhancing the breadth of the human experience of natural wonder. In physics, the Michelson–Morley experiment confirms that the speed of light is independent of motion. In technology, the year of 1887 saw the advent of the telescopic fishing rod, the punched card calculator, the wind turbine, the fire escape, the contact lens, and the world's first automatic rifle. And finally, in fantastic literature, Barcelona's Enrique Gaspar publishes *El Anacronópete*, the first work of fiction to feature a time machine. (The original Greek idea of time had a double identity of *Kairos* and *chronos*. *Kairos* suggested a moment of time, in which something special happens. *Chronos* was focused on measured and mechanical time. Little wonder that this machine age brought mechanistic time to the fore. *Chronos* became king. Time travel was born.)

But the most important thing that Watson says of Sherlock's science knowledge base is its imperfections. And maybe chief among these is Sherlock's ignorance of the Copernican Theory. The Copernican Theory is the name given to the astronomical model developed by Nicolaus Copernicus and published in 1543. Your first thought may be, *what does a theory published almost three hundred and fifty years before Sherlock's first appearance in print have to do with deductive science?* Well, as it happens, quite a lot. Copernicus's theory was that the Sun sat at the center of the Universe, and *not* the Earth. Rather, the Earth was a planet, like all the others, all of which orbited around the Sun.

So what? Well, the Copernican Revolution was a revolution of ideas, a transformation in our very conception of the Universe and our relation to it. Many thinkers, including famous American philosopher of science Thomas Kuhn, regard Copernican Theory to have kick-started the scientific revolution itself, which in turn led to Sherlock's contemporary world.

Sherlock and Watson have the following exchange on the topic of the Copernican Theory:

> **Watson:** "But the Solar System!"
> **Sherlock:** "What the deuce is it to me? You say that we go round the Sun. If we went round the Moon it would not make a pennyworth of difference to me or to my work . . . You appear to be astonished. Now that I do know it I shall do my best to forget it."
> **Watson:** "To forget it!"

Watson then says he was on the point of asking Sherlock what his work might be, and not long later Sherlock confesses, "I have a trade of my own. I suppose I am the only one in the world. I'm a consulting detective, if you can understand what that is."

The Deduction Diamond: Meaning

Finally, in using the Deduction Diamond, it's important to consider the possible *meaning* of Conan Doyle's quoted text; what's it all about? It's one of the first portrayals of Sherlock's *modus operandi*, and the narrowness of the focus of Sherlock's inquiring mind. The importance of Copernican Theory to the scientific revolution itself is *exactly* why Conan Doyle picked it. It sits as an exemplar of Sherlock's myopic vision as a consulting detective in that he probably chose to ignore Copernican Theory for the greener pastures of science more relevant to his art. As Sherlock explains to Watson:

> You see, I consider that a man's brain originally is like a little empty attic, and you have to stock it with such furniture as you choose. A fool takes in all the lumber of every sort that he comes across, so that the knowledge which might be useful to him gets crowded out, or at best is jumbled up with a lot of other things so that he has a difficulty in laying his hands

upon it. Now the skillful workman is very careful indeed as to what he takes into his brain-attic. He will have nothing but the tools which may help him in doing his work, but of these he has a large assortment, and all in the most perfect order. It is a mistake to think that that little room has elastic walls and can distend to any extent. Depend upon it there comes a time when for every addition of knowledge you forget something that you knew before. It is of the highest importance, therefore, not to have useless facts elbowing out the useful ones.

Sherlock is clearly exaggerating. Of course the Solar System and Copernican Theory make a difference to him and his work. Not just on a general level, with regard to the philosophy of science, but also to the particulars of detection. For example, the laws and properties of physics are of vital import to a consulting detective, and we can be sure that Sherlock has them down to an exact science. But he doesn't worry about the broader aspects, in the moment, when those properties are not immediately relevant. Still, his point is a valid and useful one.

In the business of decision making, Sherlock is saying it's too easy to get distracted if there's simply too much clutter in the attic of one's mind. No matter the myriad facts at one's disposal, nor how capacious and chasmic one's memory might be, and we are led to believe Sherlock's is truly huge, it's all useless unless one knows what to use and what to ignore—the hallmark of a good thinker. (We see this selection of relevant facts perfectly exemplified in the BBC's *Sherlock*. For example, in the episode "The Hounds of Baskerville," Sherlock goes to his "mind palace." Watson describes the mind palace as a memory technique with a mental map; you plot a map, where the location doesn't have to be a real place, and then you deposit memories there that theoretically you can never forget, all you have to do is find your way back to it. We then see a superb graphic sequence of Sherlock sifting through memories and concepts, some he discards, others he keeps for the case afoot.)

But there's a very real way in which Sherlock is quite wrong. And that's the idea that there is no room for a broad base of general knowledge. In the real world, one never knows quite what tools and facts you will need at any one time—it's best to be prepared. (The writers of *Sherlock* were aware of this. The storyline of the *Sherlock* episode "The Great Game" is somewhat derived from *The Adventure of the Bruce-Partington Plans*, makes some references to *The Five Orange Pips*, and borrows from other Conan Doyle works. It's also notable for its clever use of general astronomy knowledge. The plot partly revolves around the so-called Van Buren Supernova, an exploding star which only appeared in the sky in 1858, so could not have been painted by Dutch master Vermeer in the 1640s. Thus, the Vermeer is a forgery.)

Let's now take a trip back through time to the late nineteenth century. The rapidly expanding Empire has spread from Britain like the giant web of a mechanized beast. At the very center of this huge machine, the heart of the circulating veins and capillaries of commerce, sits the black macadamed streets of Victorian London. In our first adventure, *An Adventure in the Illimitable Ocean*, we shall focus on the *context* aspect of the Deduction Diamond to help answer the question: what role did science play in developing London as the dark heart of England's machine age? We shall find that London itself is an equally fascinating and shadowy character for Conan Doyle. Shrouded in fog with terrible crimes taking place under the cover of darkness, the city of Jack the Ripper and of terrifying underworlds, Sherlock is the man of science who faces the thriving underworlds of crime, engaged in a struggle to the death.

CHAPTER ONE

AN ADVENTURE IN THE ILLIMITABLE OCEAN

. . . in which we engage in social science, and wonder what kind of world conjured the London of 1887, and see the evolution of the great city through the eyes of Sherlock himself.

If you look from a distance, you observe a sea of roofs, and have no more knowledge of the dark streams of people than of denizens of some unknown ocean. But the city is always a heaving and restless place, with its own torrents and billows, its foam and spray. The sound of its streets is like the murmur from a sea shell and in the great fogs of the past the citizens believed themselves to be lying on the floor of the ocean . . . London goes beyond any boundary or convention. It contains every wish or word ever spoken, every action or gesture ever made, every harsh or noble statement ever expressed. It is illimitable. It is Infinite London.

—Peter Ackroyd, *London: The Biography* (2000)

"Dirty old river, must you keep rolling/Flowing into the night?"

—The Kinks, "Waterloo Sunset" (1967)

Sherlock's London, the Illimitable Dark Ocean

London, the year 1887. It's an autumn morning as the fog starts slowly to lift off the liquid history that is the Thames. No longer a "silver streaming" river, as Spenser lyrically labeled it in the days of Elizabeth, and already on its way to being biologically dead, as the Natural History Museum would declare it exactly seventy years from now. For centuries the city's slickers have sailed on its surface, docked, and traded on its shores, and built their homes not far from its waters.

Llundein. Lunden. Londinium. The Great Wen. The Big Smoke. Now a metropolis of five million souls. The dust and din and steam of a city that has swelled from its ancient origins to a huge urban sprawl. Known now as the greatest city on Earth, the capital of Empire, hub of wealth and trade, an immense global cauldron into which the world pours.

From the vantage point of the Golden Gallery, the cobblestones and narrow, twisting lanes are lost in the lingering fog. But the taller monuments of London stand like sentinels of Empire above the brume. Imagine the huge translucent face of a gentleman's pocket watch set down on this scene, with the watch's center wheel hole mounted on the ball and cross of Saint Paul's dome. The river is sat at six o'clock. At eight o'clock but on this, the north, side of the winding river, are Big Ben and the Houses of Parliament. At nine-thirty, facing the Thames, is the center of Sherlock's London operations at 221b Baker Street, a street that stands above the subterranean Baker Street station, one of the original stations of the world's first underground railway. Baker Street is a road that runs up to Regent's Park and London Zoo, the world's third-oldest zoo, which opened its doors to the public almost sixty years before.

At three o'clock and proximal stands the Bow Church, and distal lurks the trapezoidal shape of the Royal Exchange. In the direction of Bow and the Exchange, hidden now beneath the pea-souper, sat the tangle of medieval streets devoured by the Great Fire of 1666. A wide arc of the old wall gates that date from Roman times, running from the Tower of London at three o'clock in the east, through Aldgate at two-thirty, Bishopsgate at two o'clock, Moorgate at one-thirty, Cripplegate at one o'clock, and Ludgate at nine, marks the boundary of the conflagration. In less than a square mile, over thirteen thousand homes and almost ninety churches were destroyed. The Royal Exchange and Saint Paul's itself had also burnt down.

The English poet John Milton once wrote of the "vast unbounded Deep" of the cosmos, unveiled by the telescope, " . . . a dark Illimitable ocean, without bound, Without dimension; where length, breadth, and height, And time, and place are lost." So it is with earthly London in 1887, where the Ripper streets of Whitechapel will witness the first of those infamous murders.

London's Cacophony

Just as sight is trammeled by the murk, the sounds of London are also dampened and distorted by the fog. The city is already awash with noise. Imagine the cacophony. Pigs grunt, ravens croak, tills ring, dogs bark, saws sing. The costermongers cry their wares on Ludgate Hill. Cab drivers yell conversations from opposite sides of the road. Paper boys shout out the news. "Read about Andrian the dog-faced man!" "Spring-heeled Jack spied in Clapham!" And soon, "The ripper strikes again." Drunks sing and holler up and down the streets and squares. And the mellifluous ado of the inescapable musicians, as Charles Dickens put it, "brazen performers on brazen instruments, beaters of drums, grinders of organs, bangers of banjos, clashers of cymbals, worriers of fiddles, and bellowers of ballads."

By the middle of the century and around the time Sherlock was born, the tens of thousands of horses that drove Hansom cabs and omnibuses had met their match in the machines of the railways. Whoops of mudlarkers (kids between eight and fifteen, or the robust elderly) boasted finds from the muck of the river. Sherlock believed the banks of the Thames to be a rich source of ancient artifacts. He knew the accumulation of centuries of flotsam and jetsam eroded daily to reveal the city's past. He knew from his Baker Street Irregulars that mudlarkers were scavengers in muck. The old English idiom, "where there's muck there's brass," was true for those poor souls scouring the river's shore and scraping a subsistence living. For being a mudlark was a fate decided by poverty and lack of skills. But among the muck and waste, avoiding the corpses of cats, dogs, and fellow humans, and dodging the broken glass left and littered on the shore, a mudlarker worth his mettle might find a small fortune.

A wild boar tusk, perhaps, found on the foreshore where high status townhouses once stood, the meat of the beast destined for the dinner tables of the rich. Or a microlith, a tiny stone tool made from flint or chert dating back to three thousand or thirty thousand years ago. Or George III halfpennies, pins, Georgian buttons, bullets, lead weights, decorative brass, daggers, swords, or a small "Frozen Charlotte," a piece of porcelain that was chilled and placed into a teacup to cool a lady's tea. All evidence of London's past. Liquid history indeed.

(Mudlarking wasn't the only "fruit" enjoyed by the poor, willing to do anything to keep hungry mouths fed. Dog poop collectors sold their "wares" to tanners who used it to dye leather (and the tanners themselves had to acquire the bloody hides from butcher shops and then allow them to soak in lime pits). Sewer hunters, known as "toshers," were forced to get by with what "treasure" they could find while sifting through waste in the sewers. Leech collectors (mostly women) waded into water with bare legs to lure the leeches for delivery. Chimney sweep was a job done largely by boys as young as four years old. Grave robbers, also known as "resurrectionists," was an occupation so bad that, when there weren't enough cadavers to rely on for anatomical research, the robbers started committing the murders themselves. And finally, rat catchers, as London was flooded with rats for the rich and poor alike—but only the poor ran the very high risk of getting bitten by the rodents and contracting disease.

Sherlock's London: Jekyll and Hyde

In *Strange Case of Dr. Jekyll and Mr. Hyde*, the scientist, Dr. Henry Jekyll, a well-made man of fifty with a "slyish cast," creates a chemical serum in an attempt to hide his hidden evil, only to make matters worse. His alter ego, Mr. Edward Hyde, grows in power. (The hysteria surrounding the Ripper serial murders would reach such a fever pitch that actors who merely played killers on stage would be under suspicion of murder, and the *Jekyll and Hyde* production at the Lyceum Theatre would be forced to shut down.) In Christian theology, Lucifer's fall from grace is down to his refusal to see that he is a created being, with a dual nature.

And here is old London town with its duality of nature, created through progress in science and technology. The city's divisions of social class, its few filthy rich and masses of miserably poor, and along with that the fundamental dichotomy of the age—outward respectability and inward lust. London, a city of public probity and private vice. Just as disease spreads unseen, so these gaslit streets disguise their own dark vectors. Crime is commonplace. Pickpocketing and burglary. Violent affray and calculated murder. Child prostitution and opium dens. With such a dark underbelly, London would always need a consulting detective.

This megalopolis is at a unique point in its evolution. Never before in the history of this city was less regard shown for the great mass of its citizenry. At least Charles Dickens and Friedrich Engels were two Victorian city dwellers who had warned of the war on the poor. Half a century back before this day, Dickens had *Oliver Twist* published as a serial. The tale of *Twist* told of child labor and domestic violence and exposed the cruel treatment of the many orphans in Sherlock's city. And Friedrich Engels's book, *The Condition of the Working Class in England*, was written during his stay between 1842 and 1844 in Manchester, the city at the heart of the Industrial Revolution which birthed this machine age.

It's a weird idea, a whole city full of the poor, injured, and maimed. Yet, in quite large part, that's exactly what London is. In the early years of this very century, London still had some small-city traits. It was compact. A little like Londinium of old—capital of Britain during Roman rule when the city was contained within the old, gated ways—in the early 1800s you could still cast a small circle over the cluster of rooftops and bound the burg within the circle's sweep.

In those days, the margins of the city still boasted strawberry fields at Hackney and Hammersmith. But these strawberry fields were not forever. That smaller city still had the dark arts of cock-fights, dogfights, and public executions. Those were more eccentric times that were still only partly lit by gas, where linkboys bearing lights would usher night revelers home from the caves of harmony, and the day was full of delights such as pleasure gardens and cheap window theaters.

But then London became Sherlock's city. The city of machine time and the steam engine. The home of natural philosophy, where electromagnetic science was discovered and declared. A city whose liminal expansion on all points of the compass announced to the globe that Britain had become the first urbanized society in the world. London was the center of mass production, and the impersonal forces of capital demanded a vast army of clerks and pen pushers. In this sense, London made its citizens into slaves of the machine.

Sherlock's Dark City

London became a metropolis of light and shade. A city packed to blackness, whose climate created hieroglyphics written in soot and smoke upon its ancient stones. A city whose population would swell to seven million souls by the time of the Great War. And so London became darker. The spectrum of the city ran down a gradient from bright to dark. The uniforms of its worker bees went from bold colors to the bible black of the frockcoat and the stovepipe hat. In architecture, it was the same. The grace and color of the early century's Georgian buildings were eclipsed by the later neo-Gothic piles of Victorian Empire.

Sherlock's dark city emerged more massive, centrally controlled, and organized. It had also become far more anonymous and less human. A megalopolis in which crime could thrive. There were around half a million dwellings, enough housing to run a ring around the entire island of Britain. Every eight minutes, someone in Sherlock's city died. Every five minutes, some poor soul was born. The Big Smoke now bragged forty thousand costermongers, sellers of goods from a handcart in the city's twisted streets, and one hundred thousand winter tramps. The swollen city could crow more Irish than Dublin itself, and more Catholics than lived in that other Eternal City of Rome. London had over twenty thousand pubs, serving over half a million drinkers. And where there's drink, there's the devil.

Even Engels had confessed, "A town such as London, where a man may wander for hours together without reaching the beginning of the end . . . is a strange thing." The indescribable shock of this strange new kind of city meant that Engels was reduced to repeating images of enormity, "the immense tangle of streets," "hundreds of thousands of all classes," "endless lines of vehicles," "countless ships," "hundreds of steamers," "hundreds and thousands of alleys and courts." And Thomas Carlyle described London as "the heart of all the universe," a savage wilderness, without pity or restraint; the heart of Empire, and the heart of darkness.

That darkness had been a long time brewing. Henry Fielding had written, over a century ago now, that "the immense number of lanes, alleys, courts and bye-places" that Engels would later describe, appeared as "a vast [primal] wood or forest in which the thief may harbor with as great security as wild beasts do in the deserts of Arabia or Africa." There was neither watch nor ward to police the dark districts where criminals lurked. Desert streets within a buried city's maze of stone, a wilderness in which London itself was a primeval force where the baser instincts of humans were allowed free expression. Sherlock's dark city of Empire was also a city of savages.

Whitechapel was home to the worst of London's slums. The district was known for its tangled rat runs of alleyways and passages, high in human density and crime. Life in the Ripper Streets was grim. Over half its almost eighty thousand inhabitants lived in poverty, working up to eighteen hours a day for an Ebenezer wage. The slightest of errors could cost them dearly: a Scrooge-like docking of a day's wage, or a loss of livelihood. Work was barely any better than the lot of a slave, the factories full of dangerous fumes, the buildings covered in a film of filth, or even built over cesspits.

Home life in the Ripper Streets was no better. Many lived in common lodging houses, with rat-infested rooms crammed to the rafters so that landlords could make as much profit as possible. In such a climate, crime was a convention rather than an exception. Males might take solace in female company, providing landlords with further profit by setting up some lodging houses as brothels. Whitechapel's worst area was known as "the wicked Quarter Mile." The district included Thrawl Street, Flower Street, and Dean Street, with rooms no more than 8' x 8' that would house twenty people who cooked, ate, slept, and worked in the same space. When the Ripper murders hit, some police thought that Jack's route led to the vicinity of Flower and Dean Street, while others thought the Quarter Mile should be the epicenter of their manhunt. Jack the Ripper. One day soon, Sherlock might muse, men will look back and say that Jack gave birth to the twentieth century.

The Birth of Science and Capital

Such is Victorian London as a relatively unsung character on Sherlock's stage. Sherlock mapped the ways in which his London had become the tainted jewel in this workshop of the world, in this machine age. How had Britain become the first urbanized society in the world and London its center of mass production? Men of science like Sherlock put the answer down to Britain's status as a cauldron of science and technology and the works that it conjured.

The scientific revolution made leaps and bounds in Britain thanks to the likes of Isaac Newton. While Newton's huge sway lay mostly in physics and mathematics, his "system of the world," his paradigm of a cosmos governed by knowable laws, nonetheless laid the basis for much Enlightenment thinking.

Newtonianism became a prestigious intellectual program which applied Newton's principles to many other avenues of inquiry, including Sherlock's. But the most immediate effect was in the field of politics and economics. First, through Newton's philosopher friend John Locke and his successor David Hume, Newtonianism took the form of a skepticism of authority and a belief in *laissez-faire* that were to reduce the prestige of religion and subvert respect for an allegedly divine order of society. Then through Voltaire, who introduced Newton's ideas to the French, Newtonianism led to the program of the French Revolution.

It's easy to see why Sherlock should be so well-acquainted with Copernican Theory. The theory put the fear of science into the Church. After all, once the radical pioneers of science tired of the stars, they might just as easily turn their telescopes around to point at the Church, and the established order of their antiquated system, which is exactly what eventually happened. The Copernican Theory had led directly to Newton.

The Newtonian spirit broke with the fixed forms of feudalism and faith, and the more ancient tradition of slave-owning from the Classical world. In politics and in science, this meant human ingenuity exploded into previously closed fields. The rise of science and capital had a three-fold foundation: *people, ideas,* and *applications.* In terms of *people,* Newton, like Sherlock, had been a polymath. In Newton's case, not only a mathematician, but also an astronomer, an optician, and a mechanic. For years, like Sherlock, Newton also worked on chemistry. And Newton's nemesis, Robert Hooke, not only worked in the same fields as Newton, but also in physiology, in microscopy, and found great esteem by performing over half of the architectural surveys after the Great Fire of 1666 in Sherlock's city. As a result, Britain's early *virtuosi* had a more unitary picture of science than at any time since. They, like Sherlock, a latter-day virtuoso, saw science as a whole, rather than reduce it to its disparate parts.

The second factor of the rise of science and capital in Sherlock's country was the role of *ideas*, especially the weight given to a mathematical kind of philosophy. Here the British had borrowed from the Arabic, the Hindu, the Chinese, and especially the Greeks. The obsession with mathematical thinking explained why the *virtuosi* had huge successes in astronomy and mechanics but made little early progress in chemistry and biology. And so mathematics became a weapon of the Industrial Revolution. Its wielders viewed the world as a series of sums, of additions and subtractions, of the mere balancing of books. The monetary difference in, say, buying in a cheap market and selling in a dear one. The cash difference between production cost and selling price. Or the difference between investment and return. To the champions of this rational method, even politics and morals could be subjected to simple calculations. If happiness was an object of government policy, then every person's pleasure could be calculated as a quantity. And so could their pain. Deduct the total pain from the total pleasure and the net result of the human equation should be happiness. The government that delivered the greatest happiness, for the rich, was the best. This queasy accountancy of humanity was satirized by Charles Dickens in his *A Christmas Carol*, when Scrooge says, "This is the even-handed dealing of the world! There is nothing on which it is so hard as poverty; and there is nothing it professes to condemn with such severity as the pursuit of wealth!" Yet, satire notwithstanding, capital would create its debit and credit balances of humanity, just like in business, and poverty would be excused.

The third factor in the three-fold foundation of the rise of science and capital Sherlock knew was *applications*. This was the most characteristic principle of the new science: its concern with the technical issues of the day. Its one great success, of course, was navigation. Britain's control of the seaways opened up the world to Empire. And what an Empire it was. Sherlock lived in a century now known as *Pax Britannica*, "British Peace." Modeled after *Pax Romana*, these were the golden years, at least for some. A period of relative peace between the Great Powers.

The British Empire became the global power and adopted the role of a "global policeman," just as Sherlock became the ultimate authority on crime in his city. From 1815 and for another twenty-seven years until the Great War, this was Britain's "imperial century." During these days, around ten million square miles of territory, 24 percent of the Earth's total land area and roughly four hundred million people, 23 percent of the world population, were added to the British Empire. At the peak of its power, it was described as the Empire on which the Sun never sets, as the Sun was always shining on at least one of its territories. Such was the power of science in the pay of capital. (Critics of the day spun an alternate explanation: The Sun never sets on the Empire, because God in all his glory doesn't trust the British in the dark!)

Industrial Revolution

What got the Industrial Revolution started? Or is the better question why did the Industrial Revolution happen first in Britain? And why did it happen at this particular time? Of course, there are many necessary conditions, but what was the trigger that set the Industrial Revolution off at a particular time and place?

Newtonianism was the first factor, and an economy that was becoming more free market. Then, of course, Britain was a Protestant country, and the protestant work ethic benefited the economy. Many scientists and inventors were members of religious minorities who were not allowed to hold positions in many professions because you had to belong to the Church of England. Even other kinds of protestants, known as dissenters, were not allowed to hold such positions or even to go to the best universities. And so, in Britain, dissenters often became the leaders of the Industrial Revolution.

Britain also had a ready economy. The home market was freer, and abroad, Britain had the growing market provided by the British Empire. Profits from the slave trade, admittedly prohibited by Britain in 1809, provided capital for investments in transportation and new technology. Slave-owning planters and merchants who dealt in slaves and slave produce were among the richest people in eighteenth century Britain. Such profits helped endow All Souls College, Oxford, with a luxurious library, to build a bevy of banks, including Barclays, and to finance the experiments of James Watt, inventor of the first really efficient Newtonian philosophical engine, known now as the *steam* engine. But Britain's Industrial Revolution wasn't the result of a single invention. No, it was down to technological progress in disparate fields, coming together like an orchestra responding to a conductor. And that conductor, if it were anyone, was Isaac Newton.

What enabled the Revolution to keep going? The progress was powered on coal. Britain had heaps of coal, so it didn't run out of energy as population and industry grew, but not everywhere was as lucky as Britain. (China, too, had huge reserves of coal. And in the 1600s it was also on the brink of an industrial revolution. But it was not so lucky with its geography. The problem was, China's coal lay way up in the north of the country. And the great cities of China's Empire lay on the south coast. Between the coalfields and the coast lay the mighty Yellow River. And that meant having to ride the impossible rapids of the River. So, the only way to get the coal to the great coastal cities was to carry it, overland, a distance of hundreds of miles. As a result, the price of the coal doubled every twenty-five miles. The Empire was economically cut off from its coal, and China's industrial revolution was greatly delayed.) Britain had an inland waterway system of canals that were used to transport huge amounts of heavy produce because roads simply couldn't handle such weights, and the vehicles needed to move the produce didn't yet exist. These canals were crucial to keeping the Revolution going.

The key component of Britain's economic revolution was the development of factories, which hadn't really existed before this time. The factories couldn't have happened without good transportation creating larger markets, and better transportation couldn't have existed without the growth of the iron industry, which couldn't have grown without the steam engine. Sherlock and his fellow citizens were living during a special time. These were days marked as a new starting point for technology. Tech was changing so swiftly that people's lives were revolutionized by science within their lifetimes, rather than over many generations.

Science brought light to Sherlock's city. From the mid-1700s to the early 1800s, the Industrial Revolution meant some of London's labyrinthine streets were lit at last. Parliament had given the London and Westminster Gas Light and Coke Company, the first gas company in the world, a charter to light the city's streets in 1812. By the end of the next year, Westminster Bridge was the first in the city to be lit by gas.

But science also brought smog. Along with the factories came a new concern: polluted air. There was a good reason William Blake had called them "dark Satanic Mills . . . in England's green and pleasant Land." Factory chimneys burned coal and belched out thick black smoke. The sheer amount of suspended particulate matter doubled over the 1700s and 1800s. This problem led to smog, that infamous mixture of London smoke and fog. The famous black brick walls of 10 Downing Street are painted but they originally became that color due to the amount of soot that clung to it and other London buildings.

Most 'Orrible Crimes

Around three-quarters of the recorded crime carried out in Victorian London was petty theft. Violent crimes made up only around one in ten cases, and murder was relatively rare, yet the chattering classes of London were much more anxious about grisly crimes than the theft. Those living in the West End thought there was a crime wave that needed to be stopped, but this belief was greatly inflated by the many "penny dreadfuls," the cheap newspapers that were filled with detailed descriptions of "'orrible crimes." Victorian London's fascination with murder and murderers and such a popular interest in the macabre would soon be fueled into a conflagration by the Ripper murders. The thing about Jack was this: He lived so long in the public memory for a reason. And that's because people feel that Jack was a product of the fabric of social and economic reality of the time.

The Baker Street Irregulars, who run errands and track down intel for Sherlock, are a group of poor or homeless children who live on the city's streets. Sherlock recalls that, around forty years ago, the social reformer Lord Ashley, the "Poor Man's Earl," had reported more than thirty thousand "naked, filthy, roaming lawless, and deserted children" in and around Sherlock's city.

Yet they were all-eyes-and-ears. Vital intelligence agents for Sherlock's operation. He often introduced them as the Baker Street division of the detective police force. The Irregulars dwelt in London's underbelly, down among the city of savages and slaves, in those dark districts where few "respectable" souls strayed. The teeming streets of the wicked Quarter Mile, what would soon be called Jack's Whitechapel, provided new and greater opportunities. Especially for the pickpockets.

Pickpockets did very well in Sherlock's London. Public executions provided particularly rich pickings. London hangings could attract crowds of more than a fifth of a million souls. Just as Dickens wrote in *Oliver Twist* about Fagin, Sikes, and the child pickpockets of Saffron Hill, the city's rookeries were homes to organized gangs of pickpockets. A rookery is what the locals called a city slum occupied by the poor, criminals, and prostitutes. Rookeries were those overcrowded areas with low-quality housing and no sanitation where local factories such as coal plants and gasholders polluted the rookery air. The dark side of scientific progress.

A political specter was also haunting London. The specter of the uprising of the poor and the working class. Before this century, many riots, revolutions, and political protests, including treason and civil war, had been carried out by the middle and upper classes. But the mid-1800s saw the working class, who were still without the vote, using riots and protests as the only way to show the government their unhappiness. Chartists and Luddites. Swing Riots and the Anti-Corn Law League. Crimes, capital and petty.

In some ways, much theft was simply cold logic. Distress due to poverty gives the worker only the choice of starving slowly, killing himself quickly, or taking what he needs where he finds it—in plain English, stealing. And it is not surprising that the majority prefer to steal rather than starve to death or commit suicide.

The year is 1887. The British Empire celebrates Queen Victoria's Golden Jubilee, marking the fiftieth year of her reign. Buffalo Bill's *Wild West Show* brings the legend of notorious gunslingers, lion-hearted lawmen, and dangerous outlaws from the fringe of civilization to its center in Sherlock's London. Elsewhere, construction of the iron assembly of the Eiffel Tower starts in Paris, a city which is also witnessing this year the Dutch poverty-stricken painter Vincent van Gogh begin a set of still-life paintings called

Sunflowers, of which an oil painting in the second set would be sold a century later and make history as the first modern work of art to fetch a multimillion-dollar price.

Science created the very fabric of Sherlock's urbanized megalopolis. Newton's revolution led ultimately to the steam engine and the Industrial Revolution. All light and dark, Jekyll and Hyde. All brilliant but broken. And Sherlock's calling, as the world's first consulting detective in the world's greatest city, capital of the world's first urbanized society, was to use science to uncover the city's caliginous crimes.

CHAPTER TWO

AN ADVENTURE IN FOUR FINGERPRINTS

... in which we flirt with the psychological science of the four personality factors which fed into the creation of the scientific character of Sherlock Holmes.

"Half a capital and half a country town, the whole city leads a double existence; it has long trances of the one and flashes of the other; like the king of the Black Isles, it is half alive and half a monumental marble."
—Robert Louis Stevenson, *Edinburgh: Picturesque Notes* (1879)

"If London was an alien city, Edinburgh was another planet."
—Jess Walter, *Beautiful Ruins* (2012)

Who is Sherlock Holmes?

It's been almost one hundred and forty years since 1887, the year in which Arthur Conan Doyle first created his detective character Sherlock Holmes in the novel *A Study in Scarlet*, published in *Beeton's Christmas Annual*. Today, Sherlock Holmes has a fair claim to being the most immediately and globally recognizable fictional character in English literature. The Inverness cape. The deerstalker hat. The calabash pipe. His best friend and housemate Dr. John Watson. And his archnemesis Moriarty, a worthy adversary. The nefarious Professor Moriarty, the Napoleon of crime. The evil genius and prince of darkness. Sherlock's inverted double, as brilliant as Sherlock himself, and with whom Sherlock has a titanic confrontation. All have become part of popular culture consciousness.

But who *is* Sherlock Holmes? One part of Sherlock's universal appeal is that he is a brilliant but broken character. Sure, he's something of a science superhero. But he doesn't always solve his cases. And he has an infamous drug dependence. Sherlock was a cocaine addict. His preferred method of consuming cocaine is injecting what he calls the 7 percent solution. But the very fact that he is a drug addict makes him an incredibly powerful antihero, which both gives him his power and also renders him a liability. Sociopathic. Scientific. Genius. In Sherlock, Conan Doyle anticipated the future—a complex and flawed character so worshiped that some people forget that he never existed.

The one undeniable characteristic of Sherlock's continuing charm and popularity is his frequent and ostentatious use of science and the scientific method. Conan Doyle said that he purposely gave Sherlock "an immense fund of exact knowledge to draw upon in consequence of his previous scientific education." That's another way in which Conan Doyle made Sherlock a different kind of detective. Not that there had been many detectives before 1887. The first modern detective story is thought to have been Edgar Allan Poe's *The Murders in the Rue Morgue*, published in 1841.

In Mary Shelley's most famous fiction, Dr. Victor Frankenstein rejects the dark arts of old-world alchemists and turns to face a more scientific future. Victor creates a beautiful but broken creature using cadaver spares from charnel houses. Similarly, in creating his most famous fiction, Dr. Conan Doyle rejects the dark arts of old detectives such as Poe's Auguste Dupin, who doesn't believe in scientific and analytical methods, but compares solving a crime to gambling with cards or chess. Instead, Conan Doyle makes Sherlock a virtuoso of science.

What fingerprints are to be found on the simulacrum that is Sherlock Holmes? And from what clay was he cast? Let's look at the four personality factors which fed into the creation of the scientific character of Sherlock Holmes: Conan Doyle himself, Edgar Allan

Poe, Dr. Joseph Bell (Conan Doyle's mentor in medical school), and that leading light of all brilliant but broken characters, Isaac Newton.

Fingerprint One: Sir Arthur Conan Doyle

Arthur Conan Doyle was born in a year auspicious for science: 1859 (May 22 to be precise). Later that year, November 24, the world also witnessed the publication of one of its most famous books: Charles Darwin's *The Origin of Species*. Darwin's book injected a new lifeblood into science. The theory of evolution inspired not just scientists like Conan Doyle but also the radical, anticlerical wing in politics and its agenda of laissez-faire. Darwin's theory became the definitive philosophy of this machine age of the nineteenth century.

Conan Doyle, like Darwin, was educated at the highly respected University of Edinburgh Medical School. Medicine had been taught at Edinburgh since the beginning of the sixteenth century. In Doyle's day, a medical education consisted mostly of lectures, with very little hands-on training. Nonetheless, when Conan Doyle entered medical school, the required courses included anatomy, physiology, medical chemistry, Materia medica (drugs), morbid anatomy (pathology), surgery, midwifery, therapeutics, gynecology, children's diseases, vaccination, teeth, mental diseases, and hygiene. Quite a grounding! By the end of his schooling, Conan Doyle declared that he had become agnostic, a term itself coined only a few years earlier by scientist Thomas Henry Huxley, also known as Darwin's Bulldog for his advocacy of Darwin's theory of evolution.

The University of Edinburgh was a radical institution by the standards of British universities in the nineteenth century. Oxford and Cambridge were Anglican universities; that is, they were related to the Church of England. A student could not graduate from either without swearing allegiance to Anglican articles of faith. The science taught at these universities was informed by what was known as natural theology. And the canonical text was William

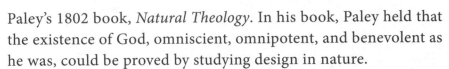

Paley's 1802 book, *Natural Theology*. In his book, Paley held that the existence of God, omniscient, omnipotent, and benevolent as he was, could be proved by studying design in nature.

In natural theology, knowing nature meant knowing what The Almighty had in mind when he created the cosmos. Teachings such as Paley's also held a political and social message. And natural theology was used to defend the status quo. Like the organic and inorganic worlds, the social world too was divinely designed and condoned.

Thus, British science in the first half of the nineteenth century was dominated by the religious beliefs of Oxford and Cambridge dons. Consider just three renowned Oxbridge examples. First, the geologist and paleontologist William Buckland who was also the Dean of Westminster. It was Buckland who wrote the first full account of a fossil dinosaur, which he named *Megalosaurus*. But, for many years, Buckland was also a fluvial geologist who believed in a global deluge during the time of Noah. Second, Astronomer John Herschel made many contributions to the science of photography. He named seven moons of Saturn and four moons of Uranus, discovered by his father Sir William Herschel, but John Herschel also opposed evolutionary theory, and later became a theologian. Finally, the English polymath William Whewell is noted for coining some well-known modern words in the English scientific lexicon, including scientist, physicist, linguistics, catastrophism, uniformitarianism, and astigmatism. But Whewell was also a creationist who held that God created the Universe in accordance with certain Divine Ideas gifted to Man, and the more we idealize scientific facts, the more difficult it will be to deny God's existence.

Contrast all the above with the University of Edinburgh where Darwin studied. Religious nonconformists who wanted a university education had to leave England entirely. Many, like Darwin, chose the University of Edinburgh. The University was a seat of radical evolutionary thought. It was open to new ideas in anatomical and biological thinking from Europe. And many of the University's radicals rejected natural theology, with its associated hierarchical view of nature. Rather, they saw the natural world, and society, evolve and progress through education, cooperation, emancipation, technology, and democracy. The lesson of French naturalist Jean-Baptiste Lamarck was paradigmatic. That a creature could, through its own endeavors, transmute into a higher being, and without the help of gods, appealed to evolutionists and progressives who dismissed the status quo.

So, when Conan Doyle enrolled at the University of Edinburgh Medical School, he was signing up to a university that had preserved its progressive status. The old English institutions of Oxford and Cambridge were bastions of opposition to the new evolutionary theories, as they were still essentially largely Anglican institutions. But the Scottish University of Edinburgh was far more open to the new ideas.

Edinburgh also had a long tradition of outstanding practitioners. As part of his medical studies, between 1876 and 1881, Conan Doyle's teachers included Charles Wyville Thomson, William Rutherford, and Joseph Bell. Thomson, a natural history specialist, had recently returned from a three-and-a-half-year journey at sea, where he studied the oceans aboard the *HMS Challenger*. Conan Doyle's physiology professor, William Rutherford, became the role model for the explorer Professor Challenger in Doyle's novel *The Lost World*. But Scottish surgeon and lecturer Dr. Joseph Bell was Conan Doyle's most influential teacher. Conan Doyle described Bell as "the most notable of the characters whom I met." Joseph Bell was a pioneering forensic pathologist. And Bell's clinical deductions made him something of a Victorian version of Gregory House, MD. Conan Doyle was dazzled by Bell's deductions about patients. So much so that he added similar scenes in his tales of Sherlock Holmes.

After receiving his Edinburgh-based education, Conan Doyle attended a family meeting in London in 1882 to discuss his future.

At the meeting, Conan Doyle told them how far he had moved from his Catholic heritage. His uncles, Richard, Henry, and James possessed money and power. They could help Conan Doyle get up the career ladder in medicine. They felt he had potential. And, once he had settled on a base for his practice, the rich uncles could use their sway with Catholic families to help Conan Doyle find prospective patients. But even though he knew it might cost him their backing, Conan Doyle nonetheless confessed to his uncles that he was an agnostic. This revelation not only shocked his uncles but also led to an enduring rupture with them.

Conan Doyle spent a stint as a ship's physician from February to August of 1880. Serving aboard the *Hope* on a voyage to the Eastern Arctic, Conan Doyle spent time in the company of the sailors and was impressed by how they had educated themselves. Doyle wrote to his mother in February 1880, "the chief engineer came up from the coal hole last night and engaged me upon Darwinism, in the moonlight on deck." Clearly, Darwin's theory of evolution was not just a controversial topic for the chattering classes in Oxford, Cambridge, and Edinburgh. It was also being debated on the decks of British ships.

Sherlock's world is a Darwinian one. And the dynamic of Victorian science and culture that influenced Conan Doyle's outlook centers on a key group of Victorian scientists: Thomas Henry Huxley (Darwin's Bulldog, whom we have already met); John Tyndall, the Irish physicist who first divined the connection between atmospheric CO_2 and what is now known as the greenhouse effect; and Herbert Spencer, the philosopher of evolution, coiner of the phrase "survival of the fittest," and considered the single most famous European intellectual in the closing decades of the nineteenth century.

The scientific development of Sherlock's creator owes so much to his careful study of Darwin and his disciples. Conan Doyle's account in his autobiography is of a "constant struggle" between his faith and "new knowledge that came to me both from my reading and from my studies." He identified those thinkers who most challenged his faith in his autobiography *Memories and Adventures*: "It is to be remembered, that those were the years when Huxley, Tyndall, Darwin, Herbert Spencer and John Stuart Mill were our chief philosophers, and that even the man in the street felt the strong sweeping current of their thought, while to the young student, eager and impressionable, it was overwhelming."

The seat of Dr. Doyle's loss of faith, which he declared to his uncles at that fateful family meeting in 1882, can be tracked back to his time at Edinburgh University's Medical School. True, Conan Doyle's favorite professor, Joseph Bell, was a pious man. But no evidence has been traced to suggest that Bell's influence negated the materialistic climate at Edinburgh. Bell's pious philosophy does not seem to have crept into his classroom. Dr. Bell's diagnostic technique, the methodology that so intrigued Conan Doyle (and, about which, more later) seems to have been totally secular in nature.

In short, at Edinburgh University, Conan Doyle enjoyed an ambience rich in radical and materialistic ideas, the kind of climate for which Edinburgh had been famous since the early 1800s. And it was the Darwinians Huxley, Tyndall, and Spencer who had most shaped Conan Doyle's early scientific thinking.

These Darwinians, Spencer, Tyndall, and Huxley, had all met in London during the 1850s. (Coincidentally, many Sherlock scholars believe his birthday to be January 6, 1854.) In those days, the Darwinians were promising young scientists, trying to make a mark for themselves. It wasn't easy. All three came from outside the Oxbridge elite. But they found that they shared lots in common. They all had humble middle-class roots. And, as we have seen, whereas the establishment's scientific elite were part Anglican theologians, these three Darwinians wanted to revolutionize science by secularizing nature, making science a profession, and advancing expertise.

It's plain to see the seeds of Sherlock's early character development in these aims of the Darwinians. Conan Doyle's portrayal of his great detective certainly seemed to secularize nature. In the 1891 story *The Five Orange Pips*, Sherlock tells Watson that the two chief qualities needed for a detective to be an "ideal reasoner" were a highly evolved deductive mind and a command of relevant scientific facts. Conan Doyle also promoted science as a profession; Sherlock was, after all, a trailblazer as the world's first consulting detective. And many of Dr. Doyle's tales advanced Sherlock's expertise.

Hundreds of thousands of Victorian readers were transfixed by the Sherlock adventures, especially his unerring ability to solve crimes through his heightened expertise and powers of reasoning. As Sherlock puts it in *Five Orange Pips*:

> To carry the art, however, to its highest pitch, it is necessary that the reasoner should be able to utilize all the facts which have come to his knowledge, and this in itself implies . . . a possession of all knowledge, which, even in these days of free education and encyclopedias, is a somewhat rare accomplishment. It is not so impossible, however, that a man should possess all knowledge which is likely to be useful to him in his work, and this I have endeavored in my case to do.

Other examples from Sherlock tales suggest that Doyle's approach was based on his careful study of Darwin and his disciples. But the tales' relationship to the aims of secularizing nature, making science a profession, and advancing expertise is a complex one. In *The Adventure of the Blanched Soldier*, Sherlock is our narrator in the absence of Watson, and he muses that his method "is but systematized common sense." But in *The Adventure of the Lion's*

Mane, Sherlock says that he holds "a vast store of out-of-the-way knowledge without scientific system." In this way, Doyle's tales set up Sherlock as a professional in the consulting detective sense, but as an idealized amateur and self-taught polymath in terms of science.

Plenty of tales pit Sherlock against villains for whom specialization has a perverting influence. Think of Professor James Moriarty, who first gains recognition at the age of only twenty-one for writing "a treatise upon the Binomial Theorem" which led to his being awarded the Mathematical Chair at one of England's smaller universities. Think also of Professor Presbury in *The Adventure of the Creeping Man*, a renowned physiologist at a great English university. And think particularly of Culverton Smith in *The Adventure of the Dying Detective*, a Sumatran plantation owner who becomes an expert in Asiatic diseases. But Smith uses his expertise for ill. First, he murders his nephew. Then, he tries to murder Sherlock. Unforgivable.

Thomas Huxley had a natty name for the vision of a liberated science: scientific naturalism. The position of the Darwinians was that true science had no room for a divine being; the job of the scientist is to stick to the study of observable causes and effects in nature. You can see the influence on Dr. Doyle's development of Sherlock clearly here. The Darwinians became the avant-garde of an important cadre of intellectuals, many of them fellow scientists, who canvassed strongly for a reappraisal of science's position in society in the second half of the nineteenth century. The peak of their power within British science was between the 1860s and the 1880s, when Conan Doyle was developing his Sherlock stories, and when the Darwinians dominated scientific societies such as the British Association for the Advancement of Science and the Royal Society.

The Darwinians weren't just interested in revolutionizing scientific philosophy and institutions. They were also evangelical about transforming British culture. For example, Huxley, who for around thirty years was evolution's most potent champion and for many "*the* premier advocate of science in the nineteenth century [for] the whole English-speaking world," also organized public lectures for workers, themselves looking for a new, liberating science. Little wonder that Conan Doyle ended up debating Darwin with sailors on the decks of British ships. The Darwinians proposed a new vision of humanity, nature, and society based on the theories and methods of empirical science, especially evolutionary biology. They tried to conjure a new scientific worldview for an evolutionary age that was losing its ties to biblical literalism and religious dogma, which naturally brought them into conflict with the Church, in the same way that Conan Doyle was brought into conflict with his family. The Darwinians claimed that they represented the best intellectual leadership for this modern machine age, this industrialized Britain. And what an exemplar of that very idea Conan Doyle's Sherlock Holmes would prove to be.

There's little doubt that Dr. Doyle was inspired by the Darwinians. They provided an alternative to the Christian view of contemporary life, and a critique of theology that would have found an echo within him after his time at Jesuit schools. In addition, the evolutionary Darwinian outlook would have been, in a very real way, a scientifically based alternate faith for a lapsed Catholic, though it did not fulfill him for the rest of his life. The Darwinians were wise enough to realize that they couldn't simply reject the Christian creed out of hand. They knew that many people were looking for new answers, a new scientific vision to supersede the old faith. One that presented a hopeful and brighter future. After all, they knew, more than most, that science was now the real driving force behind all genuine progress in this new machine age. The paradigm of the Darwinians would have been hugely seductive to Conan Doyle in his creation of Sherlock Holmes.

In fact, we can identify three stages of Doyle's relationship with the Darwinians. The first of those was as a young man when he embraced materialist and scientific thinking and rejected the idealism of the religious worldview. This allowed Doyle rational explanations of the world and shaped a relatively iconoclastic view both of the world and of humanity within it. Later, Doyle demurred from this radical scientific stance, focusing rather on Darwin and Wallace's roles as masculine explorers and virile progenitors of scientific ideas. Finally, in his later years, Doyle embraced spiritualism as a greater truth, viewing materialists as cases stuck in arrested development. This descent into spiritualism is witnessed in a piece of Doyle's autobiographical fiction, 1895's *The Stark Munro Letters*, where Doyle writes "Is it not glorious to think that evolution is still living and acting—that if we have an anthropoid ape as an ancestor, we may have archangels for our posterity?" From the same *Letters*, Doyle's ideas mutate into a more reactionary form: "Nature, still working on the lines of evolution, strengthens the race in . . . the killing off and extinction of those who are morally weak. This is accomplished by drink and immorality." The Sherlock tales belong to the middle phase of Doyle's progression.

Dr. Doyle

Conan Doyle started a medical practice of his own in the Southsea suburb of Portsmouth in 1882. Here he compensated for its slow start and growth by writing stories in his spare time. After several months of advanced study in Vienna and Paris, Conan Doyle then opened a new practice in London in 1891, continuing to write all the while, in what he called his all too abundant free time. According to legend, not a single patient ever attended Conan Doyle's London surgery.

Luckily, Dr. Doyle was also a gifted storyteller. And the Grub Street presses in London were revolutionizing the novel. The 1800s saw the rise of speculative science in fiction for the growing book-buying market. For example, in 1818 Mary Shelley's *Frankenstein* was published. Hers was a voice of dissent in response to the double-edged sword of technology and change, whereas the outlook of the *Voyages Extraordinaires* of Jules Verne saw nothing but progressive expansion for science and capital. When Darwin's *Origin of Species* was published, writers became obsessed with a technologically evolving society. Edward Bulwer-Lytton's 1871 novel *The Coming Race* foretold the race of the future, and satirist Samuel Butler anticipated machine intelligence in his farsighted 1872 novel *Erewhon*. Toward the end of the century, speculative science novels were being written for those who wished to change the world. One only has to think about the way in which H. G. Wells explored social Darwinism in his 1895 novel *The Time Machine* and *The War of the Worlds* three years later.

It was the perfect time for a fictional champion of science. Conan Doyle already published several stories, beginning in 1879, so he resolved to write a detective novel with a difference. His template was Edgar Allan Poe's detective, Auguste Dupin, but Conan Doyle's creation was a supersleuth the likes of which the world had not yet witnessed. The intelligence of Sherlock Holmes would be so superior that he could solve mysteries that left lesser mortals in befuddlement. Happenstance, so key to lesser crime stories, would be consigned to oblivion. Science would be king.

Before the nineteenth century, there wasn't much evidence in literature of the world unveiled by science, perhaps understandably. Poetry had little to do with the laws of physics. But for the Romantic Movement, the likes of Mary Shelley, a dialogue with science became a key concern. Witness also William Wordsworth's interest in science:

> If the labors of men of science should ever create any material revolution . . . in our condition . . . the poet will sleep then no more than at present, but he will be ready to follow the

steps of the man of science, not only in those general indirect effects, but he will be at his side, carrying sensation into the midst of the objects of the science itself. The remotest discoveries of the Chemist . . . will be as proper objects of the Poet's art as any upon which it can be employed.

Trying to best express the taste, the feel, the human meaning of scientific discoveries is just what Conan Doyle intended to do with his new detective fiction. To show how the scientific method can create a material revolution in the question of solving crime.

There hadn't been many fictional scientist forerunners upon which to base the character of Sherlock. Let's take just two example predecessors, divided into radical and conforming trends. The radical trend began with Mary Shelley's *Frankenstein*, the 1818 novel that became a seminal voice of contradiction on the role of science in society. *Frankenstein* is a tale of science gone astray. Victor Frankenstein is the Faust of science. Victor's crazy dream is unlimited power through science. But there is horror in the new responsibility facing the scientist. *Frankenstein* calls for vigilance in the practice of the new philosophy. It warns of the primal urges of power and control in all creations of technology. And Mary Shelley's book became a potent metaphor for the powerlessness of the scientist. In contrast, consider Jules Verne's 1864 classic, *Journey to the Center of the Earth*. The scientist in Verne's famous novel is Professor Lidenbrock, who discovers a runic cryptogram which is a key to traveling to the Earth's interior. Verne presents his novel's paradigm: Nature is a cipher to be cracked, and the scientist is the hero to do the job. From the outset, Professor Lidenbrock is portrayed as an avatar of science, locked in a lethal struggle with nature.

The stage was set for Sherlock Holmes's first appearance in print. And we have established that Sherlock's world is Darwinian. Dr. Doyle discards teleological explanations. But he still finds meaning and purpose in the natural world. Like Dickens and Poe, Conan Doyle's evolving Sherlockian worldview is secular and naturalistic. Sherlock's paradigm has no need of gods or the conservative perspective of nineteenth-century biblical literalism.

Fingerprint One and a Half: Conan Doyle's London

Another crucial innovation of Conan Doyle's is his portrayal of late-Victorian London. London is part of Sherlock's DNA, and Sherlock a crucial codon within the city's genome. With Doyle begins that notable achievement of crime fiction: to reimagine cities at crucial times in their evolution. Raymond Chandler's booming Los Angeles in *The Big Sleep* and *The Long Goodbye*, for example. The characteristic Scandinavian melancholia of Henning Mankell's Ystad in the Kurt Wallander novels. Or the independent and criminal Edinburgh of Ian Rankin. In these works, the city is central to the story. But all roads lead back to Sherlock. There can be little doubt that a word association game using the words "detective" and "city" would most often, and globally, conjure one response: Sherlock and London. Conan Doyle's tales depict, as Watson says in *The Adventure of the Resident Patient*, a great city whose "ever-changing kaleidoscope of life . . . flows through Fleet Street and the Strand."

Difficulties stood in the way of Conan Doyle's detective/city development. Doyle had limited knowledge of London as he grew up, and studied in Edinburgh. But the decision to place his detective in London was due to the influence of two Parisian sleuths, detectives Dupin and Lecoq, belonging to writers Edgar Allan Poe and Emile Gaboriau respectively. Both detectives get brief mentions early in *A Study in Scarlet*. Conan Doyle was also influenced by his artist/illustrator uncles John and Richard, both of whom lived and worked in London. Indeed, in 1874, at the tender age of fifteen, Conan Doyle visited his uncles for three weeks. They showed him the sights, including Madame Tussaud's, which was then located on Baker Street. Conan Doyle wrote to his mother that he was captivated "by the room of Horrors, and the images of the murderers." Once back at school, Doyle delighted in reporting that he had enjoyed his "three weeks in London immensely. I saw everything and went everywhere. In one walk I thoroughly saw Saint Paul's, Westminster Abbey and bridge, houses of Parliament—The Tower–Temple Bar, the Guild Hall and other places of interest."

Conan Doyle's choice of Baker Street as Sherlock's base was inspired. According to Michael F. Harrison's *In the Footsteps of Sherlock Holmes*, by 1880 the area had transformed from "high-class residential to upper-middle-class-residential and commercial." And, as Baker Street had an underground station on the new Metropolitan Railway, it had quickly become a major interchange for commuters traveling from the rapidly expanding west into the beating heart of Sherlock's city. The Baker Street area was a typical market of archetypal *Strand* readers that George Newnes would capture on launching his magazine in 1890.

Baker Street was the perfect place for the world's first consulting detective and problem solver. So was the idea of situating Sherlock's stories in *The Strand Magazine*. *The Strand* aimed to target middle-class urban and suburban readers. It offered both fiction and nonfiction journalism on national and international themes. The name of the magazine was vital. *The Strand* is a major thoroughfare in the city of Westminster. It's only a mile in length and runs from Trafalgar Square eastward to Temple Bar, where the road becomes Fleet Street in the City of London. It links the twin towers of British power—the money-obsessed City of London and the Empire-obsessed domain of Westminster. In Sherlock's day, *The Strand* also had ten theaters, around one hundred cafes and public houses, and a close link to modern media, as Fleet Street was where the London and national newspapers were situated. The first issue of *The Strand* featured a ten-page article, *The Story of the Strand*, which begins, "The Strand is a great deal more than London's most ancient and historic street: it is in many regards the most interesting street in the world."

In *The Adventure of the Norwood Builder,* Sherlock holds that "to the scientific student of the higher criminal world, no capital in Europe offered the advantages which London then possessed." But, for Sherlock, those advantages were nothing in the absence of a worthy opponent, when London became "a singularly uninteresting city."

Sherlock's intimate knowledge of his city's streets becomes a weapon in his crime-solving arsenal. In *The Sign of Four*, Watson declares that, in contrast with his own "limited knowledge of London," Sherlock's was profound and never at fault in identifying the city's streets, even in thick fog, insinuating that Sherlock's geography and his detective radar worked in phase.

There is an amusing sequence in Guy Ritchie's 2009 movie adaptation, *Sherlock Holmes*, where Sherlock is blindfolded and taken to a secret London location. Despite the blindfold, Sherlock knows he's been taken to the home of Lord Rotherham, who apologizes for the abduction and says, "I'm sure it's quite a mystery as to where you are." Sherlock's reply is delicious:

> As to where I am, I was, admittedly, lost for a moment, between Charing Cross and Holborn, but I was saved by the bread shop on Saffron Hill. The only baker to use a certain French glaze on their loaves—a Brittany sage. After that, the carriage forked left, then right, and then the telltale bump at the Fleet Conduit . . . As to the mystery, the only mystery is why you bothered to blindfold me at all.

This uncanny ability is used to assert Sherlock's authority. For instance, in *The Adventure of the Empty House*, Watson brags that "Holmes's knowledge of the byways of London was extraordinary," and he links this once more with his crime-fighting genius: "I knew not what wild beast we were about to hunt down in the dark jungle of criminal London." But perhaps the best example of Sherlock's spatial knowledge of his city occurs in *A Case of Identity*, with this incredibly voyeuristic passage:

> "My dear fellow," said Sherlock Holmes as we sat on either side of the fire in his lodgings at Baker Street, "life is infinitely stranger than anything which the mind of man could invent. We would not dare to conceive the things which are really mere commonplaces of existence. If we could fly out of that window hand in hand, hover over this great city, gently remove the roofs, and peep in at the queer things which are going on, the strange coincidences, the plannings, the cross-purposes, and wonderful chains of events, working through generations, and leading to the most outré results, it would make all fiction with its conventionality and foreseen conclusions most stale and unprofitable."

In this book's introduction, we considered the Deduction Diamond as a method for looking at Conan Doyle's Sherlock tales. They are one way in which the reader can obtain a reliable and reusable formula for coming to sound conclusions about Doyle's works. The four points of the Deduction Diamond, which all relate to one another and to Doyle's texts, were effects, techniques, context, and meaning. Taken together, they can lead to satisfying interpretations about the Sherlock tales. If we use the Deduction Diamond to analyze the quote above from *A Case of Identity*, we draw some interesting conclusions about the context of Conan Doyle's words, which give us a new insight as to the passage's possible meaning.

The vision that Conan Doyle presents in tales like *A Case of Identity* is one of Sherlock as a science virtuoso, an all-seeing agent of control over the crimes of London. This vision bears an interesting comparison with a panopticon prison. The panopticon, a name derived from the Greek word for "all-seeing," is associated with English social reformer, Jeremy Bentham. The panopticon penitentiary was based upon an idea of Jeremy's younger brother, Samuel, who while working in Russia hit upon the "central inspection principle," which would facilitate the training and supervision of unskilled workers by experienced craftsmen. Jeremy came to adapt this principle for his proposed prison, an "Inspection House" envisaged as a circular building with the prisoners' cells arranged around the outer wall and the central point dominated by an inspection tower.

Essentially, Bentham's idea of the panopticon prison's design was that of a system of control, an institution that would allow all prisoners to be observed by a single overseer, and without the inmates being able to tell whether or not they're being watched. Critics commented on the potentially disturbing quality of the pervasive surveillance of the panopticon, with its unaccountable jailer. After all, who guards the guard? And we might ask the same question of Sherlock's all-seeing power in Doyle's vision. Can Sherlock always be trusted?

What did Doyle mean by this implicit comparison between the all-seeing eye of Sherlock and the panopticon prison? Certainly, the panopticon presents a blueprint for institutions to exercise continuous silent control over people. Prisoners, or potential criminals in Sherlock's case, know that they are subject to surveillance, but never exactly sure when the gaze might be directed toward them. And so, they regulate and temper their own behavior, assuming they are *always* under surveillance, like being watched by Big Brother in George Orwell's dystopian novel, *Nineteen Eighty-Four*.

Doyle's portrayal of his detective is that of a mythological culture hero in Sherlock's ability to oversee and read his city. The Victorian city in the machine age was a new phenomenon. It was a rapidly expanding beast which newly necessitated and confronted police competence to map and crack the cases of crime it generated. No matter how uncontrollable the city's problems, and however impossible to solve by one all-seeing and all-knowing virtuoso, Doyle nonetheless gave his readers the sense that Sherlock *could* do so. All Sherlock need do was to "mount a high tower in his mind," as Charles Dickens said of Inspector Bucket, for Sherlock to survey his city from a central and superior panoptical vantage point. In taking Sherlock out of the city in some stories, Doyle has his detective stand for a panoptical surveillance of not just the city, but of the English countryside, too. Indeed, Sherlock proves to be more suspicious of country crime than its city's counterpart. In *The Adventure of the Copper Beeches*, for instance, Sherlock's concern is the unknowable nature of the English landscape, as he admonishes Watson to "[t]hink of the deeds of hellish cruelty, the hidden wickedness which may go on, year in, year out, in such places, and none the wiser."

The main outcome that arises from this all-seeing condition is that it helps to construct a different feeling for those being surveilled. Whether a prisoner or potential criminal, those policed are always visible, while those policing (Sherlock) stay so obscure that they remain unverifiable yet omnipresent. Such a system favored by Doyle would presumably lead to self-regulation due to the internalization and different subjectivity of this policing. People would be badgered into good behavior. Thus, Sherlock's "super vision" seems to mirror Bentham's idealized panoptic policing. Sherlock is centralized, but at the same time he needs to remain obscure and remote. While in an ideal and efficient panoptic state everyone polices themselves, the occurrence of crime shows that the system in fact is realistically doomed to failure. Victorian crime was so very often based on need and want and inequality. Such distresses provide poor people only the choice of starving, suicide, or committing crime—panopticon or no panopticon. Little wonder most prefer crime, which necessitates calling upon the services of the omniscient Sherlock.

Many modern remixes of Sherlock, particularly those in film and television, seem to have something of a fetish with fog, forever drenching London's twisted lanes in vast volumes of a mad mist. Conan Doyle was a little more reserved with the stuff. Fog is a relatively rare and unstressed meteorological occurrence in Doyle's tales, save for one notable exception: *The Hound of the Baskervilles*, where Doyle has the fog lend a suitably Gothic air to the out-of-London moors.

Whereas London's fog is mostly ignored by Conan Doyle, the city's kaleidoscope of cultures is not. The kaleidoscope was a nineteenth century invention of Doyle's Scottish countryman, David Brewster. It is an optical device whose reflections produce changing patterns when the tube is in movement. And so it is with Sherlock's kaleidoscopic city, in which the color of London's culture changes as one moves through it. In *The Adventure of the Six Napoleons*, Doyle has Sherlock and Watson take something of a magical mystery tour through Victorian London, from Kennington to Stepney. Through our dynamic duo, we witness an array of contrasting spaces: "[In] rapid succession we passed through the fringe of fashionable London, hotel London, theatrical London, literary London, commercial London, and, finally, maritime London."

Thinking again about the four points of the Deduction Diamond, we might draw some interesting conclusions about Doyle's intent from this passage in *The Adventure of the Six Napoleons*. Our inseparable pair take a rather circuitous route as Conan Doyle makes their journey longer than need be. For instance, Doyle has them traverse the Thames early, in order to show off the city's kaleidoscopic culture, rather than have them travel directly through London, south of the river. What is the effect of the passage? What impressions does Doyle convey and what is the passage's meaning? Two main points occur. One, it reminds the reader that any such portrayal of London is a fictional construct. It is a selected and edited ordering of a kaleidoscope of different and competing cultures and districts. Sure, Sherlock knows the way around his city, yet any sense we might get of Sherlock's London as a whole is as eccentric as Sherlock's own esoteric filing system back at Baker Street. Two, our dynamic duo's tour underlines the fact that city space is not just for passing through, on the way to a desired destination. The journey itself is a lesson in life.

Finally, Sherlock hardly ever strayed into the Ripper streets. This last curiosity of Conan Doyle's portrayal of Sherlock's city is an interesting one. If his London is a criminally generative space, then surely the East End, the areas around Whitechapel and "the wicked Quarter Mile" dark districts where few respectable souls strayed and where poverty was profound, should be a regular haunt for a crime hunter like Sherlock? For many, Sherlock's presence is required on the Ripper streets. But it's almost as if Conan Doyle felt that Sherlock's presence had ideas of its own and sent his absence along as a proxy, with a note reading "hope you get the hint."

The Deduction Diamond might help us fathom Sherlock's absence in those Ripper streets. Consider *The Man with the Twisted Lip*. It's the sole Doyle story in which Sherlock's caseload leads to a journey, for substantial periods, into the East End, with its dangerous and threatening atmosphere. Context is all important here. The tale is a revealing study with respect to Doyle's appreciation of city space and social class. The tale begins with Watson, gallantly coming to the rescue of his wife's friend, taking a daring trek into an East End opium den. Doyle's conflation of the East End with the Far East suggests a mild xenophobia about the way in which a dangerous foreign other can violate the so-called civilized space of the West. (Perhaps Doyle forgot about the Opium Wars between 1839 and 1860. For many years, Qing China and the UK were locked in a conflict triggered by the dynasty's campaign to enforce its prohibition of opium against British and American merchants, who sold opium produced in India and Turkey to Chinese smugglers.) When Watson gets to the East End, he finds

Sherlock already at work on a case. But Doyle's description of the East End is clear: a primitive, almost diabolical space, to be found at the pit of "a steep flight of steps leading down to a black gap like the mouth of a cave."

This context helps us realize what Doyle is doing and explains why Sherlock was so seldom seen stalking the Ripper streets. Doyle's readers wanted adventures about districts far grander than their own suburban reality. After all, *The Strand Magazine* was obsessed with an emergent celebrity culture, along with accompanying tales that frequently featured the higher strata of society, cultivating a fetish among its readership with social "betters." The real context of Conan Doyle's tales is of a London where most crime was petty, the likes of mugging and pickpocketing, and occurred in the lower-class districts across the city. But the settings of the Sherlock stories belong, like the cases he cracks, to the realm of magazine-buying respectable Londoners. The tales home in on the criminal activities about which most *Strand* readers were fearful, particularly if they became rich enough to migrate to the comfortable country areas just outside the city, and maybe even the kinds of crimes they could imagine themselves capable of committing.

The Deduction Diamond helps us see that the criminal space of Sherlock's London, the city essentially invented and presented to us by Conan Doyle, is highly edited and discriminatory. (As Doyle confessed in a March 1890 letter to Joseph Stoddart, editor of *Lippincott's Monthly Magazine*: "It must amuse you to see the vast and accurate knowledge of London which I display, I worked it all out from a Post Office map.") Naturally, at times this selective edit is casual, such as those tales where we visit London districts where Watson lives and works. At other times, more than half of the tales total, in fact, the readers find themselves outside of London altogether. Nonetheless, the dominance of the megapolis remains. London, the great capital, is at the center of a giant web of crime, one whose influence is national, indeed international, making Doyle's magazine-reading citizens strongly aware of their city's huge impact on the world stage.

With the help of the Deduction Diamond, we've seen how the subsidiary fingerprint of London is a crucial factor on the character of Sherlock Holmes. Whether it's in Wandsworth or Wimpole Street, Dartmoor or Derbyshire, Norwood or Norfolk, Doyle's criminological portrayal of London seldom strays far from the concerns and ambitions of his urban readership. If Sherlock *does* leave his city, it is most often by train, an easily imagined journey into the country. And so, taken as a whole, Conan Doyle's canon is a potent body of work for reflecting the fear of crime felt by the respectable middle-class and London-based readership of *The Strand*.

Fingerprint One, Reprise

Conan Doyle's debut detective novel, *A Study in Scarlet*, was at first rejected by several publishers before being bought by Ward, Lock & Co. *A Study in Scarlet* was greeted with a lukewarm reception in Britain. But in the United States, Sherlock was immediately in business, and *A Study in Scarlet* galvanized an early and excited coterie of Sherlock fans.

Indeed, America saved Sherlock Holmes. The indifferent reaction in Britain might have meant the end of the great detective. But, in 1889, the Philadelphia-based *Lippincott's Monthly Magazine* invited Conan Doyle and the Irish literary genius Oscar Wilde to meet in London. They shared a meal at the Langham Hotel during what Conan Doyle would declare a "golden evening." Stephen Fry, who not only holds the distinction of being the youngest ever member of the Sherlock Holmes Society, but also played Oscar Wilde on stage and screen, digs deeper into that golden evening:

Each met for the first time and really liked each other. And that's the surprise. It tells you a lot about Conan Doyle, and it tells you a lot about Wilde. Conan Doyle, we think of because he's got this crisp mustache and these blue eyes and you might know, if you know anything about his life, that he was a great sportsman, as well as a great writer and a doctor. So, you think of him as tweedy and serious and butch. I think meeting a figure like Wilde, who was a true literary man, a man who had this wisp of Baudelaire and a sort of French decadence about him, I think the fact that he approved of Conan Doyle, because he was a very acute critic, emboldened Conan Doyle to write again. That, and the money, of course!

By the end of that night, the authors each agreed to write a novel. Wilde went on to write his only novel, *The Picture of Dorian Gray*, while Conan Doyle was gifted the impetus to reward the world with the rest of his Sherlock stories. His promised novel, which eventually became *The Sign of Four,* paid homage to Wilde by creating a key character, Thaddeus Sholto, after him. Like *A Study in Scarlet*, *The Sign of Four* was to become one of only four Sherlock novels.

The Sherlock floodgates had been opened. When Conan Doyle had his third Sherlock story published, *A Scandal in Bohemia*, so began the seminal set of Sherlock short stories published in *The Strand Magazine. Scandal in Bohemia* was the first of the fifty-six fantastic short stories of deduction. Sherlock Holmes, the revolutionary new virtuoso of science, hit Britain like a lightning bolt. The circulation of *The Strand* rocketed to half a million every time a Sherlock story was published. And the publisher, George Newnes, calculated that a further hundred thousand copies were sold when a Sherlock tale was printed on its pages. The trifling income of Conan Doyle, the doctor, was now dwarfed by the extravagant earnings of Conan Doyle, the author.

Conan Doyle proved to be one of *The Strand*'s most popular and prolific writers. Up to his death in 1930, there was hardly an issue which didn't contain a contribution from him. With regard to income, consider the serialization of *The Hound of the Baskervilles* in 1901–1902 which was thought to have increased *The Strand*'s circulation by around thirty thousand. And Conan Doyle was handsomely rewarded. He was being paid between £480 and £620 per episode, depending on length. In today's money, that's, very roughly, around £50,000 (or $68,000) per episode. In 1900 currency that would have bought you twenty-two horses or sixty-three head of cattle (not that Conan Doyle was in the business of making such pastoral purchases).

But Doyle became tired of Sherlock quite quickly. He contemplated Sherlock's murder as early as his eighth story. Dr. Doyle's mother was on his case, however. She simply didn't approve of Sherlock's untimely death and told her son to ditch the idea of the dead detective. Indeed, Mrs. Doyle went so far as to suggest a plot twist that her son used in *The Adventure of the Copper Beeches*, Sherlock's fourteenth tale.

Oscar Wilde is among the most quotable writers of all time. You may be familiar with quotes such as, "To love oneself is the beginning of a lifelong romance" or "There is only one thing in the world worse than being talked about, and that is *not* being talked about" or "Be yourself; everyone else is already taken." Wilde also said, "I always pass on good advice. It is the only thing to do with it. It is never of any use to oneself." And this seems to be what Conan Doyle did with the advice from his mother regarding the great detective.

So, Sherlock died his "death" because Conan Doyle felt that the detective was taking up too much of his time. Time better spent on writing "worthy" literature, historical novels and the like. But there was a deeper truth behind this excuse. Conan Doyle was finding it increasingly demanding to devise new plots for his detective fiction.

The tales became a tad formulaic. He had somewhat appropriated Poe in Sherlock's first three tales. He also used the unmarried but wealthy maiden plot in the fifth, tenth, and fourteenth tales: *A Case of Identity*, *The Speckled Band*, and *The Copper Beeches*. We also get a slew of dubious dads. The stepfather in *A Case of Identity* is pusillanimous, the stepfather in *The Speckled Band* is formidable, and the father in *The Copper Beeches* is scheming. The three tales contrast starkly in terms of quality. Sherlock fans often vote *The Speckled Band* top of polls for the fifty-six short stories, whereas *A Case of Identity* fares badly.

Looking at the fan reception of Doyle's tales more closely, we can see how the formula weighed a little heavily on his creativity. Of the fifty-six stories, it is telling to compare fan polls of the first thirty stories with the last thirty. The Sherlock canon has been rated several times, including this early set of results carried out by readers of *The Baker Street Journal* in 1959 (see on next page).

To anchor these two tables in time, recall that Sherlock and Moriarty had taken that plunge over the Falls in the twenty-sixth tale, *The Final Problem*. The next story was *The Hound of the Baskervilles*, widely considered to be one of the very best. Indeed, the next three stories after *Hound* are also well regarded so, as Doyle hits his halfway point of a full thirty stories, his writing is creatively strong. However, probably due to the formulaic nature of the tales and the pressure of writing them, the quality drops off after that halfway mark. As can be seen from the tables above, eight of the ten best tales are drawn from the first half of the canon. Only two later stories, *The Bruce-Partington Plans* and *The Six Napoleons* buck that trend, both of which, interestingly, formed the basis for episodes of the BBC's *Sherlock* (2010's *The Great Game* and 2017's *The Six Thatchers*).

The situation in the ten worst tales is the opposite. A whopping nine from this list are drawn from the second half of Doyle's canon, and eight of those are from the last dozen stories that Doyle wrote between the years 1921 and 1927. And it's not as if we are stating

Rankings of Doyle's Sherlock Tales: Ten Best

Story Name	Story Number
The Speckled Band	10
The Red-Headed League	4
The Blue Carbuncle	9
Silver Blaze	15
A Scandal in Bohemia	3
The Musgrave Ritual	20
The Bruce-Partington Plans	42
The Six Napoleons	35
The Dancing Men	30
The Empty House	28

Rankings of Doyle's Sherlock Tales: Ten Worst

Story Name	Story Number
The Mazarin Stone	49
The Veiled Lodger	59
The Yellow Face	17
The Blanched Soldier	56
The Three Gables	55
The Creeping Man	51
The Retired Colorman	58
The Lion's Mane	57
The Sussex Vampire	52
The Missing Three-Quarter	38

anything controversial here. Even Doyle was in agreement. When he listed his favorite tales in 1927, a full fifteen were early stories and a paltry four were later ones.

The situation is much the same even if we include the four Doyle novels. *The Hound of the Baskervilles* knocks *The Speckled Band* off the top spot, but the later tales still fare rather poorly. Even one of the exceptions to the ten best rule, *The Bruce-Partington Plans*, probably nudges a top slot due to the "Mycroft Effect," a welcome showing of Sherlock's ever-popular brother.

Sherlock's deductions are truly startling and unsettling at times, but there is comfort to be found in the regularity of his clientele. In the Sherlock stories, there are nineteen cases where Holmes is called upon by a law-abiding bourgeois, who has somehow bumbled into a "queer business." There are sixteen stories where a government apparatchik or member of the English aristocracy consult Sherlock on "a matter of utmost delicacy." There are ten tales where a vulnerable young woman, without friends, has "no one else to turn to." There are nine cases where a salt-of-the-earth plodding policeman finds himself out of his depth with "something of a puzzler." Finally, there are six stories in which there is either no client or the client is the villain (see Figure 2 on page 25).

That most heinous of crimes, murder, or at least manslaughter, appears thirty-seven times in the Sherlock tales. Theft or robbery appears fourteen times. In six stories, *The Sign of Four, The Boscombe Valley Mystery, The Five Orange Pips, The Gloria Scott, The Dancing Men,* and *Black Peter,* Conan Doyle uses a plot where someone returns to England only to be tracked and blackmailed or threatened.

As we saw earlier, London is undoubtedly Sherlock's city. The mythic link between Sherlock's scientific detection and the machine age space of London is so very strong. Even more so when we realize to our surprise that, of the fifty-six short stories, only twenty-one are *mostly* set in the city itself. And, of these, three tales (*The Adventure of the Greek Interpreter, The Adventure of the Final Problem,* and *The Adventure of the Illustrious Client*) are only partially located in London, being also set in outer suburbs or overseas. A total of thirty-one tales take place in the Big Smoke, based on the borders of London as they were when Conan Doyle wrote the stories. A number of cases occur in locations which are satellites of London: eight in Surrey, five in Sussex, four in Kent, three in Hampshire, two in Berkshire, Essex, Middlesex, Norfolk, and one each in Bedfordshire and Cambridge. The rest are further afield with cases in Devon, Cornwall, Yorkshire, Birmingham, Herefordshire, Camford (half Oxford, half Cambridge), and finally France and Switzerland. There are two cases with backstories in the states, but sadly Sherlock doesn't materialize in America. However, when Conan Doyle *does* situate stories in nonurban areas well outside London, there is usually some strong connection with the capital. London is omnipotent, as Watson declares in *The Sign of Four*, "the monster tentacles which the giant city was throwing out into the country" (see Figure 3 on next page).

Little wonder that in December 1900 an article by Conan Doyle appeared in *Tit-Bits* in which he writes "When I had written twenty-six stories, each involving a fresh plot, I felt it was becoming irksome, this searching for plots." And so it should be of no surprise that in *The Adventure of the Final Problem*, Dr. Doyle has Sherlock die unromantically in the arms of nemesis Moriarty, as together they cascade over the Reichenbach Falls.

The news of Sherlock's death didn't go well for Dr. Doyle. We have seen how the reception in London was severe. Not only were black armbands of mourning sported by many fans, but Conan Doyle got bags full of critical letters in the post. *The Strand*'s sales tumbled, as Sherlock and Moriarty had tumbled over the Falls. And it took a full decade before Sherlock fans learned, in 1903's *The Adventure of the Empty House*, that Sherlock had never plummeted over the Falls in the first place.

The "Great Hiatus." That name coined by Sherlockians to refer to the lost decade when Sherlock was considered dead. On Sherlock's glorious return, sales of *The Strand* soared, as did Dr.

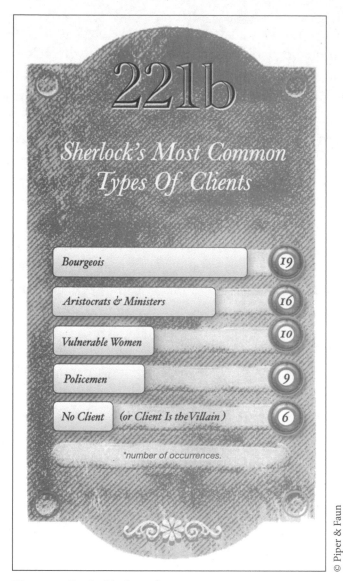

Figure 3. Sherlock's clientele.

© Piper & Faun

Doyle's bank balance. Sherlock had proved too lucrative. There was no way that Sherlock could be abandoned in Switzerland, or anywhere else for that matter. Conan Doyle may no longer be a doctor, but one patient he certainly needed to keep alive was his own creation—his pet detective.

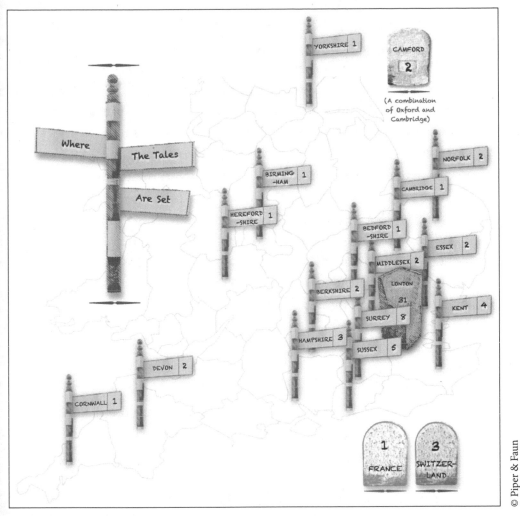

Figure 4. Where the tales are set.

© Piper & Faun

There we have it. Sir Arthur Conan Doyle, a character almost as extraordinary as his famous creation. Coming in at six feet, three inches tall and 220 pounds, Conan Doyle was an accomplished sportsman, cricket champion, prize fighter, and a matchless snooker player. He was a pioneer of, among other things, competitive downhill skiing, the life jacket, and the inflatable raft. He was an adventurer at heart, a chivalrous spirit ready for anything, and larger than life, just like Sherlock.

Finally, should we be in any doubt of Conan Doyle's scientific determination in creating Sherlock, let's turn to one last piece of evidence. It is the summer of 1927. Charles Lindbergh becomes the first man to cross the Atlantic by plane nonstop. German physicist Werner Heisenberg begins to develop quantum science's Uncertainty Principle. And Philo Farnsworth will soon capture an experimental electronic television image, of a straight line, at his laboratory in San Francisco. Progress in science and technology proceed apace.

Meanwhile, Conan Doyle has just published his final Sherlock tale. We are a mere three years shy of Doyle's death, and he records one of history's first sound-on-film interviews. How did he come to write the Sherlock tales? In answering this question, Doyle emphasizes two main factors. One, the importance of his own scientific training as a medical doctor. Two, his frustration with some of the "old-fashioned" detective fiction that he read for pleasure. In such fiction, Doyle confesses, "the detective always seemed to get at his results either by some sort of a lucky chance or fluke, or else it was quite unexplained how he got there . . . That didn't seem to me quite playing the game." So, Doyle explains, he spied an opportunity. An opening in fiction where he "began to think of turning scientific methods, as it were, onto the work of detection." Doyle then details his drawing of inspiration from the example of Joseph Bell, giving him "a new idea of the detective."

It's time we turn to the other personality "fingerprints" to be found on the simulacrum of Sherlock Holmes: Edgar Allan Poe, Dr. Joseph Bell, and Isaac Newton.

Fingerprint Two: Edgar Allan Poe

Famous for his cultivation of mystery and the macabre, Edgar Allan Poe was the first well-known American writer to earn a living through writing alone. Best known for his short stories, widely regarded as a central figure of American Romanticism, as well as a significant contributor to the emerging genre of science fiction, Poe is thought to be the inventor of the detective fiction genre. In the preface to *The Adventures of Sherlock Holmes*'s 1902 edition, Conan Doyle gave Poe credit for comprehensively inventing the classic detective fiction genre. Doyle wrote:

> Edgar Allan Poe . . . was the father of the detective tale, and covered its limits so completely that I fail to see how his followers can find any fresh ground which they can call their own. For the secret of the thinness and also of the intensity of the detective story is, that the writer is left with only one quality, that of intellectual acuteness, with which to endow his hero. Everything else is outside the picture and weakens the effect. The problem and its solution must form the theme, and the character-drawing be limited and subordinate. On this narrow path the writer must walk, and he sees the footmarks of Poe always in front of him. He is happy if he ever finds the means of breaking away and striking out on some little side-track of his own.

Poe's invention came with the publication in 1841 of his *Murders in the Rue Morgue*. The very word detective wasn't coined until 1843 and, between Poe's first foray and Conan Doyle's, there lurks a cluster of crime stories which rely far too much on conjecture, happenstance, and deathbed confessions.

Nonetheless, these tales not only act as a bridge between Poe's stories and the tales of scientific detection of Dr. Doyle, but they also provide a body of work for Conan Doyle to critique and improve upon. Thus, Conan Doyle rejuvenated detective fiction in 1887. At first, it's clear that Conan Doyle relied heavily on Poe. For instance, and from the get-go with the first Sherlock tale, *A Study in Scarlet*, Dr. Doyle borrows the idea of a cerebral detective with a confidant companion. So developed the theory that Sherlock was based on Poe's Auguste Dupin, whose Watson sidekick is an anonymous narrator.

The mark of Poe can also be detected in the second Sherlock story, *The Sign of Four*. The mysterious murderer in Poe's *Murders in the Rue Morgue* turns out to be an Orangutan, the species of great ape which diverged genetically from the other hominids, including us humans, between about nineteen and sixteen million years ago. Poe's Orangutan shimmies up a seemingly unclimbable vertical wall, as wild Orangutans move through trees by both vertical climbing and suspension. Wall scaled, the Orangutan brutally murders two women, then disappears by the same way. It was one of the very first locked-room mysteries. In *The Sign of Four*, Conan Doyle riffs on the same plot. Dr. Doyle replaces the Orangutan, height for height, with Tonga, a pygmy from the Andaman Islands, who matches the Orangutan's circus trick with an unclimbable wall and slays Sholto in another locked-room mystery.

The fingerprints of Poe can also be detected in the third Sherlock tale, *A Scandal in Bohemia*, the first of the celebrated set of short stories. Once more Conan Doyle riffs on a Poe plot. In Poe's *The Purloined Letter*, the third of the three tales featuring the famous Parisian amateur detective Auguste Dupin, the sleuth is on the stalk for a compromising letter written by the French queen. The letter, secreted in plain sight, is detected by Dupin by a cunning plan in which he causes a diversion with gunshot fire, pockets the genuine letter, and replaces it with another. Sherlock does the same in *A Scandal in Bohemia*, only the letter idea is replaced by the far more incriminating document of a photograph. (As photography was introduced to the world in 1839, it was a little too early for photographs to appear in Poe.) The incriminating photograph in question is of the king of Bohemia and Irene Adler, one of the most notable female characters in the Sherlock story series, despite appearing only in this one tale. Conan Doyle has Watson toss a smoke bomb through the window to the cry of "fire!" Agitated by the prospect of losing the photo, Adler reacts in such a way that her actions signal to Sherlock that the photo is locked away in her safe. What's more, Conan Doyle teases Poe, making out that a mere letter is nowhere as incriminating as a photo, even though it would have been cutting edge indeed for Poe to include a photograph in a story in December 1844:

> **Sherlock:** "I fail to follow your Majesty. If this young person should produce her letters for blackmailing or other purposes, how is she to prove their authenticity?"
> **King:** "There is the writing."
> **Sherlock:** "Pooh, pooh! Forgery."
> **King:** "My private note-paper."
> **Sherlock:** "Stolen."
> **King:** "My own seal."
> **Sherlock:** "Imitated."
> **King:** "My photograph."
> **Sherlock:** "Bought."
> **King:** "We were both in the photograph."
> **Sherlock:** "Oh, dear! That is very bad! Your Majesty has indeed committed an indiscretion."
> **King:** "I was mad - insane."
> **Sherlock:** "You have compromised yourself seriously."
> **King:** "I was only Crown Prince then. I was young. I am but thirty now."
> **Sherlock:** "It must be recovered."

There is even a possibility that the character of Dupin is the progenitor of Sherlock's brilliant but broken personality. Dupin has his own idiosyncrasies. Just as Conan Doyle was to do later, Poe gave Dupin the kind of peculiarities that might make his detective more memorable. But, as Dupin only features in three of Poe's tales (*Murders in the Rue Morgue, The Mystery of Marie Rogêt*, and *The Purloined Letter*), there is hardly enough story time for Poe to show some evolution in Dupin's character. Dr. Doyle, meanwhile, had sixty stories in which to (reluctantly!) develop Sherlock. But, once Sherlock Holmes became a household name, Conan Doyle had no need of character quirks, so Dr. Doyle had Dr. Watson insist Sherlock go to rehab, albeit privately, and the drug habit was ditched.

Both detectives, Sherlock and Dupin, possess dual personalities. Poe describes Dupin as having a "bi-part soul." He is both "creative and resolvent," imaginative and rational. He exhibits a "diseased intelligence," perhaps echoing Poe's divergent philosophical beliefs of idealism and materialism and reflecting the more general notion of duality and the split between the rational and irrational human virtues. Dupin's "bi-part soul" arguably provided a compelling psychological blueprint for the fictional detectives that followed, especially when amplified by Conan Doyle's Sherlock.

Holmes too shows duality. In *The Red-Headed League*, Watson compares Sherlock the consulting detective with Sherlock the musician:

> All the afternoon he sat in the stalls wrapped in the most perfect happiness, gently waving his long fingers in time to the music, while his gently smiling face and his languid dreamy eyes were as unlike those of Holmes the sleuth-hound, Holmes the relentless, keen-witted, ready-handed criminal agent, as it was possible to conceive.

We also learn that in his "singular character the dual nature alternately asserted itself." Sherlock was a galvanized soul when inspired by a consultation but, without the quickening of a case, a "self-poisoner by cocaine and tobacco" and a sufferer of intense ennui and melancholy.

This question of a dual nature is partly tied up with the idea of the antihero. Antiheroes may perform actions that are morally right, but not always for the right reasons. And they often act mainly out of self-interest. There is a longer history of antiheroes than one would maybe imagine, running from Homer's *Thersites*, *Don Quixote*, and the heroes created by the English poet Lord Byron. But Romanticism in the nineteenth century helped forge new forms of the antihero, such as the Gothic double, where the duality within a character centers on the polarity of good and evil.

In the late 1800s, this question of the dual nature of humanity was very much a fin-de-siècle art in the shadow of Darwin's relatively recent evolutionary theories. Society was still digesting his ideas. As we've seen, Robert Louis Stevenson's 1886 novella, *Strange Case of Dr. Jekyll and Mr. Hyde*, was a Gothic double—a meditation upon the duality of human nature—the inner struggle between good and evil, the Gothic at the heart of being human, the animal within the "civilized" heart. Stevenson's novella has been described as one of the best guidebooks of the Victorian era for the way in which it echoes the contemporary concern with the public and private division, the individual's sense of playing a part and the class division of London. *Jekyll and Hyde's* exploration of civilization versus barbarism was further explored in Joseph Conrad's 1899 novella, *Heart of Darkness*. While Conrad's novel is ostensibly a critique of European colonial rule in Africa, a very central idea to *Heart of Darkness* is the idea that there is little difference between "civilized people" and "savages." And, of course, by the turn of the century, Austrian psychoanalyst Sigmund Freud had become famous for developing his theory that humans have an unconscious in which sexual and aggressive impulses are in perpetual conflict for supremacy with the defenses against them.

Conan Doyle's detective tales are electrically charged with the Jekyll and Hyde polar opposition of the dark city outside and Sherlock the asocial genius withdrawn inside his apartment.

Sometimes the darkness spreads beyond the borders of Sherlock's city. As Anthony Horowitz puts it:

> What I love about the books is the way evil spreads its tentacles. So in *The Sign of Four* it's a temple in Agra in India. In *The Valley of Fear* it's the valley of Vermissa in America, or in *A Study in Scarlet* it's the plains of Utah. Places far, far, away, that nobody in London knows, yet which somehow invade London with their evil. It is not necessarily London. The books may happen there, but the evil comes from everywhere.

After Darwin, there was much interest in the way that nature seemed a stage of deceptions and tricks. Naturalists before Darwin had also noted that some animals seemed to disguise themselves as a means of survival. But this was put down to a mere minor detail in God's creation, and an opportunity was missed. Darwin's theory of evolution by natural selection, however, challenged this view. Evolutionary theory saw camouflage and disguise as a way of getting ahead in a natural world of constant flux, competition, and potential peril, if not extinction. Interestingly, both Dupin and Sherlock use the science of disguise in their detection. Dupin dons green spectacles as a disguise in *The Purloined Letter*. Likewise, Conan Doyle has Sherlock use disguises in *The Sign of Four*. Indeed, Sherlock employs disguises fourteen times in a dozen different stories.

We spoke earlier of the influence of the three Darwinians Spencer, Tyndall, and Huxley. The way in which they wanted to revolutionize science by secularizing nature, making science a profession, and advancing expertise. In both Poe and Conan Doyle, the official police force is portrayed as being inferior to the professional consultation of the experts Dupin and Sherlock. In Poe's first Dupin tale, he is disliked by the prefect. In the second tale, the prefect visits Dupin, and in the third, the prefect gifts the consultation of a case to Dupin. A similar evolution happens with Sherlock. At first, there is animosity between Sherlock and Scotland Yard. But this soon develops into an uneasy alliance, total collaboration, and finally mutual respect.

But let's not lose sight of the fact that Sherlock is primary and Dupin secondary. Famous science fiction writer Isaac Asimov had much to say about the comparison of Sherlock with Dupin. Little interest in Dupin remains, while Poe is still widely read. With Sherlock, it's the opposite. Asimov wrote that there are few societies devoted to Dupin, whereas Sherlock is such "a three dimensional living person" and so popular that many people assume Sherlock was a real-life figure. For instance, in a 2008 survey of British teenagers, 58 percent of those polled believed that Sherlock was a real individual. Multiple statues of Sherlock stand around the world, including a sculpture depicting both Sherlock and Watson that was unveiled in Moscow in 2007. A number of London streets are also still associated with Sherlock.

But perhaps the best example of Sherlock as a "three dimensional living person" happened on October 16, 2002. The world's media reported on the curious occurrence of a fictional character being awarded an honorary fellowship by one of Britain's foremost professional bodies: the Royal Society of Chemistry. True, the Society was also recognizing the hundredth anniversary of Conan Doyle's knighthood and the publication of his most famous Sherlock story, *The Hound of the Baskervilles*, but the main event was honoring Sherlock, not his creator. The honorary fellowship was conferred upon Sherlock for his use of analytical chemistry and forensic science, "the first detective to exploit chemical science as a means of detection," making him the only fictional character thus honored. The ceremony took place outside Baker Street tube station. And a gold medal was hung around the neck of the Sherlock statue there by a Fellow of the Royal Society, one Dr. John Watson.

In Japan, Sherlock became hugely popular in the 1890s as Japan was opening up to the West. He is regularly cited as a British fictional Victorian who left an enormous creative and cultural legacy in the country. A number of countries have issued stamps that bear the image and name of Sherlock Holmes, but none of Arthur Conan Doyle. Once more, most of those stamp depictions have Sherlock in the famous deerstalker hat, a creation of illustrative artists of Sherlock's tales rather than a creation of Conan Doyle himself.

As a final piece of evidence of the way in which Dr. Doyle improved upon Poe, consider the famous Sherlock quote from *The Adventure of Beryl Coronet:* "When you have eliminated the impossible, whatever remains, however improbable, must be the truth." In spirit, the quote appears to have been lifted from Poe's *Murders in the Rue Morgue,* "Now, brought to this conclusion in so unequivocal a manner as we have been, it is not for us, as rational men, to reject it on account of apparent impossibilities. It is only for us to prove that these apparent impossibilities are, in reality, not such." We should be grateful that Conan Doyle took the sometimes long-winded and dull musings of Dupin and transformed them into the crisp and succinct cleverness of Sherlock. As the English crime writer Dorothy L. Sayers put it, Dr. Doyle was "sparkling, surprising, and short."

Fingerprint Three: Dr. Joseph Bell

If we were to trust the opinion of Dr. Doyle alone, the question of the personality factors which fed into the creation of the scientific character of Sherlock Holmes would be a sole fingerprint: that of Joseph Bell. As Conan Doyle put it in May 1892, "Sherlock Holmes is the literary embodiment of a professor of medicine at Edinburgh University." So, let's take a look at this man who spent his entire medical career at Edinburgh.

Dr. Joseph Bell (1837–1911) could be variously described as a poet, a naturalist, and a sportsman. But first and foremost, Bell was a surgeon and for an entire generation he was editor of the *Edinburgh Medical Journal*. He graduated with an MD from the University of Edinburgh in 1859, the same year Darwin had his *Origin of Species* published. Bell was appointed as a demonstrator in anatomy and at the age of only twenty-six he began his own surgery classes. In due course, he was chosen as special assistant to pioneering Scottish surgeon Professor James Syme, and in 1872 Bell became senior surgeon at the Royal Infirmary of Edinburgh and later the first surgeon at the Royal Hospital for Sick Children.

Arthur Conan Doyle had to pay to sit in on the surgery classes which Bell taught at the Royal Infirmary. Bell appointed Conan Doyle as his outpatient clerk at the Royal Infirmary in 1878, which gifted him the chance to examine that for which Bell was most famous, conjuring quick diagnoses from the slightest evidence. Conan Doyle was dazzled by Bell's brilliance:

> He would look at the patient, he would hardly allow the patient to open his mouth, but he would make his diagnosis of the disease and also very often of the patient's nationality and occupation and other points entirely by his part of observation. So, naturally, I thought to myself, well, if a scientific man like Bell who's to come into the detective business you wouldn't do these things by chance, you get the thing by building it up scientifically.

Most academics agree that, at least in terms of explicit influences upon the creation of Sherlock, the application of Bell's diagnostic techniques to detective fiction is the most clear-cut.

It's worth noting again that, though Bell was a pious man, his technique was entirely secular in approach. In his 1910 article "The Romance of Medicine," Conan Doyle recollected that an education at Edinburgh Medical School was rigorously materialistic in nature. "I was educated in a materialistic age," Dr. Doyle recalled. "We looked upon mind and spirit as secretions from the brain in the same way as bile was a secretion of the liver. Brain centers explained everything . . . That was, roughly, the point of view of the more advanced spirits among us."

What form did Bell's technique take, and where did he exhibit it? The answer lies in the outpatient clinic at the infirmary that Bell conducted every Friday. To the jaw-dropped disbelief of patients and students alike, Bell would make his deductions. He was impressively adept at diagnosing not just the patient's conditions, but often their occupations, where they lived, and how they'd traveled to the clinic. When he served as Bell's outpatient clerk for these Friday sessions, Dr. Doyle was able to watch the way in which Bell analyzed minutiae and drew logical conclusions from them.

In the most Sherlockian manner, one famous example centered on a woman and her small child whom Bell had just met for the first time. Bell paraded his deductions through a set of questions:

Bell: "What sort of crossing did you have from Burntisland?" [Burntisland was a former royal burgh and parish in Fife, linked to Edinburgh by ferry.]
Woman: "It was guid." [Scottish for "good."]
Bell: "And had you a good walk up Inverleith Row?" [Inverleith Row is a prime residential district in Edinburgh.]
Woman: "Yes."
Bell: "And what did you do with the other wain?" ["wain" meaning a young child or toddler.]
Woman: "I left him with my sister in Leith."
Bell: "And would you still be working in the linoleum factory?" [Some parts of Scotland were famous for such factories, which used linseed oil to make this floor covering, commonly with a burlap or canvas backing.]
Woman: "Yes, I am."

How had Bell done it? How had he come to such stunning deductions? He had clocked the woman's Scottish accent, slightly different to Edinburgh proper, and suggestive of an out-of-town origin. The woman also had red clay on her shoes. As red clay was not found within many miles of town, except in the Royal Botanic Garden in the Inverleith district of Edinburgh, Bell deduced that the woman had taken a shortcut down Inverleith Row to arrive at the infirmary. As the child's coat was too large a fit, Bell deduced "the other wain." And, as the woman suffered from dermatitis, a common condition in local linoleum workers, Bell had accurately determined her occupation. Or, in Bell's own words, and sounding remarkably like Sherlock:

> You see, gentlemen, when she said good morning I noticed her Fife accent, and, as you know, the nearest town in Fife is Burntisland. You notice the red clay on the edges of the soles of her shoes, and the only such clay within twenty miles of Edinburgh is the Botanical Gardens. Inverleith Row borders the gardens and is her nearest way here from Leith. You observed the coat she carried over her arm is too big for the child who is with her, and therefore she set out from home with two children. Finally, she has dermatitis on the fingers of her right hand, which is peculiar to the workers of the linoleum factory at Burntisland.

Another example of Bell's brilliant deductions which so astonished Conan Doyle was Bell's snap diagnosis of a patient's condition before even examining him:

Bell: "Well, my man, you've served in the army."
Patient: "Aye, Sir."
Bell: "Not long discharged?"
Patient: "Aye, sir."

Bell: "A Highland regiment?"
Patient: "Aye, sir."
Bell: "A non-com officer?"
Patient: "Aye, sir."
Bell: "Stationed at Barbados?"
Patient: "Aye, sir."

Once more, Bell's explanation of his deductions sounds just like Sherlock: "You see, gentlemen, the man was a respectful man, but did not remove his hat. They do not in the army, but he would have learned civilian ways had he been long discharged. He has an air of authority and he is obviously Scottish. As to Barbados, his complaint is elephantiasis, which is West Indian and not British."

Bell held that the smallest traces on the body of a patient may bear silent witness to the life they lead. Though such signs may go unnoticed by others, Bell nonetheless told an interviewer in 1892 that:

> Nearly every handicraft writes its sign-manual on the hands . . . The scars of the miner differ from those of the mason. The shoemaker and the tailor are quite different. The soldier and the sailor differ in gait, though last month I had to tell a man who was a soldier that he had been a sailor in his boyhood . . . The tattoo marks on hand or arm will tell their own tale as to voyages; the ornaments on the watch chain of the successful settler will tell you where he made his money. A New Zealand squatter will not wear a gold mohur, nor an engineer on an Indian railway a Māori stone. Carry the same idea of using one's senses accurately and constantly, and you will see that many a surgical case will bring his past history, national, social, and medical, into the consulting-room as he walks in.

The secret to Bell's deductive conjuring lay in scientific method and acute observation. In his 1904 book, *The Original of Sherlock*

Holmes, Dr. Harold Emery Jones, a fellow student with Conan Doyle, quoted Joseph Bell's explanation of his methods: "Use your eyes, sir! Use your ears, use your brain, your bump of perception, and use your powers of deduction. These deductions, gentlemen, must, however, be confirmed by absolute and concrete evidence." Jones also gave account of Bell meeting a new patient and declaring to his attentive students:

> Gentlemen, a fisherman! You will notice that, though this is a very hot summer's day, the patient is wearing top-boots. When he sat on the chair they were plainly visible. No one but a sailor would wear top-boots at this season of the year. The shade of tan on his face shows him to be a coast-sailor, and not a deep-sea sailor—a sailor who makes foreign lands. His tan is that produced by one climate, a "local tan," so to speak. A knife scabbard shows beneath his coat, the kind used by fishermen in this part of the world. He is concealing a quid of tobacco in the furthest corner of his mouth and manages it very adroitly indeed, gentlemen. The summary of these deductions shows that this man is a fisherman. Further, to prove the correctness of these deductions, I notice several fish-scales adhering to his clothes and hands, while the odor of fish announced his arrival in a most marked and striking manner.

Bell's observational prowess was such that apparently immaterial facts might mean something to Bell alone and only take on greater import in due course. Over thirty years before Alexander Fleming extracted penicillin from mold in the late 1920s, Bell tutored nurses with the following stunning conclusion:

> Cultivate absolute accuracy in observation, and truthfulness in report . . . For example, children suffering from diarrhea

of a wasting type sometimes take a strong fancy for old green-moulded cheese, and devour it with best effect. Is it possible that the germs in the cheese are able to devour in their turn the *bacilli tuberculosis*?

Conan Doyle had Sherlock use similar deductive techniques when Holmes met Watson in *A Study in Scarlet*:

Sherlock: "Observation with me is second nature. You appeared to be surprised when I told you, on our first meeting, that you had come from Afghanistan."
Watson: "You were told, no doubt."
Sherlock: "Nothing of the sort. I knew you came from Afghanistan. From long habit the train of thoughts ran so swiftly through my mind, that I arrived at the conclusion without being conscious of intermediate steps. There were such steps, however. The train of reasoning ran, 'Here is a gentleman of a medical type, but with the air of a military man. Clearly an army doctor, then. He has just come from the tropics, for his face is dark, and that is not the natural tint of his skin, for his wrists are fair. He has undergone hardship and sickness, as his haggard face says clearly. His left arm has been injured. He holds it in a stiff and unnatural manner. Where in the tropics could an English army doctor have seen much hardship and got his arm wounded? Clearly in Afghanistan.' The whole train of thought did not occupy a second. I then remarked that you came from Afghanistan, and you were astonished."

Doyle has Sherlock magic up masterly Bell-like deductions in several other tales. In *The Adventure of the Greek Interpreter,* Sherlock and his brother Mycroft spar over their deductions of another military officer, with Conan Doyle suggesting it was Mycroft who had the superior mind. Another fine example is in *The Red-Headed League* when Sherlock meets a client named Jabez Wilson: "Beyond the obvious facts that he has at some time done manual labor, that he takes snuff, that he is a Freemason, that he has been in China, and that he has done a considerable amount of writing lately, I can deduce nothing else."

The BBC's flagship television series *Sherlock* starring Benedict Cumberbatch as Sherlock and Martin Freeman as Watson brought Conan Doyle firmly into the modern age. Sherlock is now a "high-functioning sociopath" who takes Bell's deduction technique to its extreme and impolite conclusion. In 2010's "A Study in Pink" when Sherlock, John Watson, and Lestrade examine the latest crime scene, the body of Jennifer Wilson, who was dressed in pink, Bell's observations are married to some very slick dialogue:

> **Sherlock:** "Okay, take this down."
> **Lestrade (tetchily):** "Just tell me what you've got."
> **Sherlock:** "I'm not gonna write it down."
> **Lestrade (angrily):** "Sherlock!"
> **John (taking out a notebook and pen):** "It's all right. I'll do it."
> **Sherlock:** "Thank you. The victim is in her early thirties. A professional person, going by her clothes; I'd guess something in the media, going by the frankly alarming shade of pink. She's travelled from Cardiff today, intending to stay in London for one night. That's obvious from the size of her suitcase."
> **Lestrade:** "Suitcase?"
> **Sherlock:** "Her suitcase, yes." (John looks around the room and frowns when he can't see a suitcase anywhere.)
> **Sherlock:** "She's been married several years, but not happily. She's had a string of lovers but none of them knew she was married."

Lestrade: "For God's sake, if you're just making this up . . ."
Sherlock (pointing down to her left hand): "Her wedding ring—look at it. It's too tight. She was thinner when she first wore it; that says married for a while. Also, there's grime in the gem setting. The rest of her jewelry's recently been cleaned; that tells you everything you need to know about the state of her marriage." (Writing in his notebook, John shakes his head with an admiring smile.)
Sherlock (down on his knees again, moving the woman's fingers to show the rings to John): "Inside of the ring is shinier than the outside—that means it's regularly removed. The only polishing it gets is when she works it off her finger but it can't be easy, so she must have a reason. Can't be for work; her nails are too long. Doesn't work with her hands, so what or rather who does she remove her ring for? Clearly not one lover; she'd never sustain the fiction of being single over time, so more likely a string of them. Simple."
John (admiringly): "Brilliant." (Sherlock looks at him in surprise.)
John (apologetically): "Sorry." (As he looks back to his notebook, Sherlock looks round almost sheepishly at Lestrade.)
Lestrade: "Cardiff?"
Sherlock (standing up again): "Obvious, isn't it?"
John: "It's not obvious to me."
Sherlock: "Dear God. What's it like inside your funny little brains? It must be so boring." (Squatting once more, he points down at the body.)
Sherlock: "Her coat: slightly damp. She's been in heavy rain in the last few hours. No rain anywhere in London until the last few minutes. Under her coat collar is damp, too. She's turned it up against the wind." (Again John shakes his head in amazement while he continues writing.)

Sherlock: "There's an umbrella in her left pocket but it's dry and unused: not just wind, strong wind—too strong to use her umbrella. We know from the suitcase that she intended to stay a night, so she must have come a decent distance but she can't have travelled more than two or three hours because her coat still hasn't dried. So, where has there been heavy rain and strong winds within the radius of that travel time?" (Standing up, he gets his phone from his pocket and shows to Lestrade the webpage he was looking at earlier, displaying today's weather for south Wales.)

Sherlock: "Cardiff."

John (grinning as he continues to make notes): "Fantastic!"

Sherlock: "D'you know you do that out loud?"

John (looking up at him): "Sorry. I'll shut up."

Thus it's hardly a shock to learn that Conan Doyle named Joseph Bell as the main fingerprint for the scientific character of Sherlock Holmes. Doyle made the assertion first in an interview in May 1892, saying that Sherlock was based on one of his medical school surgeons. In another interview the following month, he named Bell outright. And finally, on the publication of the first twelve short stories in *The Adventures of Sherlock Holmes* in October 1892, Conan Doyle dedicated the book to Joseph Bell.

Incidentally, Conan Doyle may have also been secretly aware that Joseph Bell was among many great analytical minds asked to aid in the Jack the Ripper case. Bell is said to have worked with others, possibly including police surgeon Sir Henry Duncan Littlejohn, to independently reach the same conclusion just a week before the Ripper murders ended. Though there is no record of the name of the killer Bell identified, there has been consistent speculation that the police were hiding a scandal. For example, the Ripper *may* have been James K. Steven, teacher of Prince Albert Victor.

So far, we have seen that the character of Sherlock is an amalgam of Conan Doyle himself, Edgar Allan Poe's Dupin, and Dr. Joseph Bell. Conan Doyle received a bedrock scientific education from the Darwinians, and Poe set the stage for the very idea of a cerebral detective. Poe produced the plots whose stories became themes upon which Dr. Doyle could conjure his Sherlockian variations. And Dr. Joseph Bell represented the persona upon which Conan Doyle could dress his character, the man whose mold he could shape into a detective genius.

Fingerprint Four: Sir Isaac Newton

Yet one fainter fingerprint is missing. That leading light of all brilliant but broken scientific characters, Isaac Newton. Mathematician, physicist, alchemist, and philosopher, Newton is widely recognized as one of the most brilliant men in history and is thus among the most influential science role models. Newton's work had beauty, simplicity, and elegance. His is widely thought to have been the greatest work of science ever created. Quick recap: Newton was the seventeenth century natural philosopher who first uncovered the laws of physics that govern the cosmos. He made up new branches of mathematics, conjured up the composition of light, and divined the laws of gravity and motion which hold sway across the entire Universe. Not a bad résumé.

Newton was brilliant. But in what way was he "broken?" In 1936, a huge archive of Newton's private manuscripts was put up for auction at Sotheby's in London. The papers had been kept from the public for over two centuries. One hundred lots of the manuscripts were bought by the famous British economist John Maynard Keynes. Keynes found that many of Newton's papers were written in a secret cipher. And for six years, Keynes struggled to decipher them. He hoped they would reveal the private thoughts of the founder of modern science but what the code actually revealed was another far darker side to Newton's work. For, in the manuscripts, Keynes found a Newton unknown to the rest of the world. A Newton obsessed with religion, and a purveyor of practices of heresy and the occult.

Newton's work ushered in an age, the Newtonian Age, based on the notion that all things in the cosmos were open to rational understanding. In the century or so before Conan Doyle, Newtonianism had become an illustrious intellectual program which applied Newton's principles to many other avenues of inquiry, including Sherlock's.

Whether deliberate or unconscious, there is much of Newton's personality type in Conan Doyle's creation of Sherlock's character. Consider the most monumental biography of Newton, *Never at Rest*, written by historian of science Richard S. Westfall. Here's what Westfall has to say about Newton in the preface to his imposing 1980 book, at the end of two whole decades of toil that left him wondering why he'd ever taken on so daunting a task:

> The more I have studied him, the more Newton has receded from me. It has been my privilege at various times to know a number of brilliant men, men whom I acknowledge without hesitation to be my intellectual superiors. I have never, however, met one against whom I was unwilling to measure myself so that it seemed reasonable to say that I was half as able as the person in question, or a third or a fourth, but in every case a finite fraction. The end result of my study of Newton has served to convince me that with him there is no measure. He has become for me wholly other, one of the tiny handful of supreme geniuses who have shaped the categories of the human intellect, a man not finally reducible to the criteria by which we comprehend our fellow beings. Had I known, when in youthful self-confidence I committed myself to the task, that I would end up in similar self-doubt, surely I would never have set out.

The passage is reminiscent of the way Watson might have written about Sherlock.

Keynes said that Newton was really a blend of Copernicus and Faustus, and that his alchemy was just as important to him as his physics. But the main reason that Newton entered the literature of his times, his impact even making inroads into poetry, was that he was a kind of savior who brought order to a Universe of chaos. His stature was all the heightened as he arrived at a vital time, a time when people were uncertain and even frightened about what the new Universe might bring.

The new techniques of science gave a glimpse of the very vastness of the cosmos. Earth had become an alien planet. The paradigm shift of the scientific revolution of Newton cut both ways. Not only had Copernicus made Earths of the planets, but he also brought the alien to Earth. The universe of Newton's ancestors had been small, static, and Earth-centered. It had the stamp of humanity about it. Constellations bore the names of Earthly myths and legends, and a magnificence that gave evidence of God's glory.

The *new* Universe seemed inhuman. And the further out the telescopes probed, the darker and more alien it became. The realization was terrifying to many. And yet along came Newton. A genius who believed himself to be God's emissary on Earth. A philosopher who made it his mission to find the secret ciphers of nature, hidden in both the Bible and the Greek myths, which he interpreted as encoded alchemical recipes. Newton essentially declared to the world, look, this new cosmos is not so terrifying; it can be explained. Observe my laws of motion which hold not just for your own backyard but for the furthest reaches of the Universe.

To help convince the world, Newton published his five hundred page masterpiece, the *Principia Mathematica*, which many regard to be the greatest book of science ever written. Not only was *Principia* the most magnificent work, but it was also the most all-encompassing work, and the most daring science book ever written. Newton had published a system of the world, a theory of everything.

Consider the Newtonian parallels with Sherlock. As Newton's eighteenth century feared what the new Universe might bring, Sherlock's London became the moonless jewel in the workshop of the world that was contemporary Britain. The first urbanized society, with the megalopolis of London the center of mass production of this machine age. Progress in science and technology ushered in this new industrial realm, and its dark underbelly of crime. Social critics saw London as a desert maze of stone, a wilderness in which the wicked ways of humans were allowed free expression. The chattering classes in particular viewed London as a city of savages. They were fearful of a tsunami of grisly crime that would break on the streets at any moment. This fascination with the macabre broke into hysteria with the Ripper murders. But London had its savior.

Sherlock Holmes was the Newton of crime. Like Newton, he would use the methods and techniques of science to show the machine world not only that crime doesn't pay, but that with virtuosi like himself on the trail of perpetrators, the game would always be proverbially up. The dark streets of Sherlock's London might have been conjured in the cauldron of science, but Sherlock could wield science as a weapon to right wrongs and put the dark genies back in the box.

It's possible to identify character similarities between Newton and Sherlock, too. This correspondence is gifted a new perspective when we compare the legendary life of Newton with the modern-day take on Holmes represented by the BBC production of *Sherlock*. Their dispositions, their assertive asexuality, their superior wise-guy ways, and their relationships with others all point to a similar nature.

Newton was something of a solitary figure. He immersed himself in his work, had few hobbies, and never married. Though a professor at Cambridge for almost three decades, he showed little interest in his students or in teaching. His lectures were poorly attended, and often not a single soul showed up. Rather, Newton's métier was his research, and especially the pursuit of his secret hobby in alchemy, an early form of chemistry. According to the Royal Society's Milo Keynes, Newton had Asperger's Syndrome: "The clinical features of Asperger's syndrome are, firstly, social impairment shown by poor nonverbal communication, poor empathy and failure to develop friendships; secondly, a lack of interest in communication with others; and, thirdly, an all-absorbing dominant interest and strong adherence to routine." Newton no doubt suffered from such symptoms at points during his life. But Keynes warns against an irresponsible retrofit of Newton's condition. It's far too easy to cherry pick Newton's personality and diagnose him accordingly.

Avid watchers of *Sherlock* might say the same of the show's eponymous hero. He is avowedly a "high-functioning sociopath," a seemingly isolated individual whose emotion and expressions are understated. And what we might call a "closed-off" character leads to huge feats of the creative imagination. *Sherlock* mocks the aspiration to total knowledge while at the same time, and generally speaking, presents Sherlock as having an amazing Newtonian level of knowledge, yet Sherlock's mastery is undermined by his frequently remarked-upon social ineptitude. Mary Watson declares that he knows nothing about human nature. In Newton it was mathematical physics and alchemy. In *Sherlock*'s case it was Sherlock's mind palace. Sherlock Holmes, whatever incarnation, crams an incredible amount of data into that head of his. And he has to be ready to decant some of that data as he whips up his deductions and solves the most mysterious of mysteries.

There we have it. Four main fingerprints are to be found on the simulacrum that is Sherlock Holmes. The clay from which he was cast is a potent amalgam of Conan Doyle himself, Edgar Allan Poe, Dr. Joseph Bell, and Sir Isaac Newton.

CHAPTER THREE

AN ADVENTURE IN DEDUCTION

. . . in which we look at Sherlock Holmes's deductive methods.

In solving a problem of this sort, the grand thing is to be able to reason backwards. That is a very useful accomplishment, and a very easy one, but people do not practice it much. In the every-day affairs of life it is more useful to reason forwards, and so the other comes to be neglected. There are fifty who can reason synthetically for one who can reason analytically . . . Let me see if I can make it clearer. Most people, if you describe a train of events to them, will tell you what the result would be. They can put those events together in their minds, and argue from them that something will come to pass. There are few people, however, who, if you told them a result, would be able to evolve from their own inner consciousness what the steps were which led up to that result. This power is what I mean when I talk of reasoning backwards, or analytically.

—Sir Arthur Conan Doyle, *A Study in Scarlet* (1887)

An Anatomy of Sherlock: Curious

When we think of Sherlock's skills and abilities, what first springs to mind? In *The Adventure of the Final Problem*, Watson refers to "the singular gifts by which my friend Mr. Sherlock Holmes was distinguished." In *The Sign of Four*, Watson describes him as a sharp and highly intelligent individual. At the start of that novel, Sherlock is injecting cocaine. He explains that he needs to numb his brain when there's a scarcity of cases to occupy and challenge his brilliant mind. Sherlock is clearly self-satisfied with his mental faculties, but to the point of often being dismissive of those who he thinks are less blessed.

Sherlock has a brilliant but broken mind, sharp observation skills, and great deductive powers. It's been said that Sherlock was a highly skilled forensic scientist before the field of forensic science even existed. This is true. In Sherlock's day, it wasn't called forensic science. It was usually called medical jurisprudence. And originally it was the business of physicians, who would call in various experts who knew something about, say, shoes, or fine wines, and use their knowledge. There was little organization for it at all at the Metropolitan Police, which was itself a very new organization.

What specific skills and abilities of Sherlock aided him in his work? Well, he's certainly well-versed in relevant disciplines from chemistry to anatomy to ballistics, and he integrated his vast knowledge with the relatively low-tech of the Victorian era: a magnifying glass and a ten-power microscope. But it's Sherlock's raw capacities of curiosity and deduction that must be the starting point of any consulting work.

It was Albert Einstein who once said "I have no special talents. I am only passionately curious." Sherlock Holmes was certainly an incredibly curious character. As can be learned from reading Conan Doyle's stories, or watching filmic representations of Holmes on screen, Sherlock is an antihero. He doesn't consult on cases out of some superior moral compass. No, he consults mainly due to his nature, as he relishes the thrill of curious cases.

What's happening in Sherlock's brain when a case piques his attention? To some extent, scientists are still trying to figure that out. We could say that curiosity is what galvanizes Sherlock to discover what is true about his world. Curiosity is that sensation which comes when he senses the separation between what he knows and what he wants to know. Curiosity is, if you like, Sherlock's cerebral itch, and the only way to scratch the itch is to satisfy his mind by seeking out new knowledge.

Curiosity is a basic human impulse. It's a vital skill in science, one that helps scientists make better predictions about what will happen in the future. Researchers have found that human brain chemistry changes when we get curious. The piqued interest requires the activation of brain centers which are known as the *ventral tegmental* area and *nucleus accumbens*—both centers associated with regulating the sensation of reward. Another area of the brain which is excited when we are curious is the *caudate*, which sits at an intersection between new knowledge and positive emotions, making it feel good to be curious and to learn.

Why should Sherlock be super curious compared to mere mortals? Researchers believe it has to do with genetics, age, and possibly early exposure. There are cultures that believe early exposure to specific stimuli may result in an affinity for an occupation associated with the stimuli. Sadly, not much is known of Sherlock's early life, only that he attended at least one of his country's leading universities. But one can imagine he may have been a particularly curious child. The factors of genetics and early exposure are supported by the fact that Sherlock's brother Mycroft is also super curious and possesses deductive powers exceeding even those of his younger brother, yet Sherlock claims that Mycroft is unsuitable for consulting detective work because he is unwilling to put in the physical effort needed to bring cases to their conclusions. In *The Adventure of the Greek Interpreter*, Sherlock says of Mycroft:

He has no ambition and no energy. He will not even go out of his way to verify his own solutions, and would rather be considered wrong than take the trouble to prove himself right. Again and again I have taken a problem to him, and have received an explanation which has afterwards proved to be the correct one. And yet he was absolutely incapable of working out the practical points.

If Sherlock's opinion of Mycroft is correct, we may add conscientiousness and determination to the skills that make Sherlock a fine forensic scientist.

Researchers who penned a 2011 study for *Perspectives in Psychological Science* found that curiosity is a big part of cerebral performance. According to the study, curiosity seems to be as vital as intelligence in determining how well students do in school and university. Little wonder Einstein claimed that curiosity is *more* important than intelligence. So, we can say that curiosity is not the exclusive domain of the superintelligent like Sherlock. Rather, curiosity along with other behaviors certainly inform intelligence and have a big impact on decision-making and intuition.

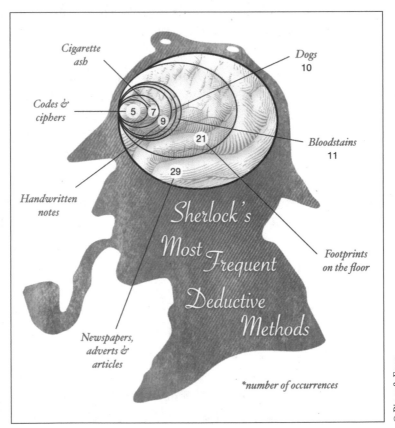

Figure 5. Sherlock's deductive methods.

An Anatomy of Sherlock: Deductive

The other main factor that Sherlock brings to each case is his deductive prowess. Sherlock's brilliance, including his curiosity and powers of deduction, is the most crucial factor in his consultations on crime. We meet Sherlock's deductive prowess from the get-go in *A Study in Scarlet*. Young Stamford introduces Sherlock to John Watson and the very first utterance Sherlock speaks in Watson's direction is, "How are you? You have been in Afghanistan, I perceive." Watson replies with, "How on earth did you know that?" The rest is history!

Another example of Sherlock's deductive acuity comes in *A Scandal in Bohemia*. Watson, now married, calls in on Sherlock at Baker Street. Sherlock proceeds to give Watson the Joseph Bell treatment; he can see that Watson has begun to practice medicine again, that he has been getting rained on lately, and even that he has a servant (female) who is not only careless but also clumsy. The reader is left in no doubt as to the validity of Sherlock's conclusions when Watson replies by suggesting that Sherlock would have been burned at the stake for such deductions a few centuries back.

In the BBC's *Sherlock* rendering of *A Scandal in Bohemia,* an episode entitled "A Scandal in Belgravia," Sherlock's deductive talents are showcased to great comedic effect in a snapshot sequence which shows a stream of visitors to 221B over a period of many weeks who consult with Sherlock and try convincing him to take up their case:

> **Man:** "My wife seems to be spending a very long time at the office."
> **Sherlock:** "Boring . . ."
> **Woman:** "I think my husband might be having an affair."
> **Sherlock:** "Yes . . ."
> **Creepy guy (holding a funeral urn):** "She's not my real aunt. She's been replaced—I know she has. I know human ash."

Sherlock (pointing to the door): "Leave . . ."
Businessman (sitting on the dining chair while two aides stand behind him): "We are prepared to offer any sum of money you care to mention for the recovery of these files."
Sherlock: "Boring."

And so on.

With regard to the Deduction Diamond and context we know that Doyle's tutor Joseph Bell was influential upon Doyle's development of Sherlock's mental acuity. Indeed, Sherlock waxes like Joseph Bell on a number of other occasions. Consider the case of *The Adventure of the Norwood Builder* in which Sherlock, like Bell, declares to the total stranger John McFarlane, "You mentioned your name as if I should recognize it, but I assure you, that beyond the obvious facts that you are a bachelor, a solicitor, a Freemason, and an asthmatic, I know nothing whatever about you."

Sherlock and Mycroft's powers are on display together when Conan Doyle had the Holmes brothers play out a deductive duel in *The Adventure of the Greek Interpreter*:

Mycroft: "Look at these two men who are coming towards us."
Sherlock: "The billiard-marker and the other?"
Mycroft: Precisely. "What do you make of the other?"
Sherlock: "An old soldier, I perceive."
Mycroft: "And very recently discharged."
Sherlock: "Served in India, I see."
Mycroft: "And a non-commissioned officer."
Sherlock: "Royal Artillery, I fancy."
Mycroft: "And a widower."
Sherlock: "But with a child."
Mycroft: "Children, my dear boy, children."
Watson: "Come, this is a little too much."

This scene is one of those that supports the idea that it was Mycroft who had the superior mind.

Nonetheless, and again like his real-life counterpart Joseph Bell, Sherlock is clearly able to make incredible deductions from the most ordinary of objects. In *The Hound of the Baskervilles*, Sherlock and John Watson duel over the analysis of a walking stick of one Dr. Mortimer, who had left the stick at Baker Street after a visit the previous night. Neither Sherlock nor Watson know the first thing about said Dr. Mortimer, and so this afforded them the opportunity to go head-to-head on the deductive process.

Watson goes first. Using the deductive methods of Sherlock, Watson's analysis of the stick leads to several conclusions. First, that Dr. Mortimer is a successful man of some considerable years. How does Watson read this from the stick? The fact that the stick was a gift from "members of C.C.H.," according to an engraving on the stick's silver band, suggests that the stick's owner is well esteemed. Watson concludes that Mortimer is a rural practitioner and that C.C.H. refers to the local hunt club, whose grateful members have presented the stick to their friend and local doctor in appreciation for his years of service.

It's Sherlock's turn next, and his reading of the stick is quite different. He declares, "I am afraid, my dear Watson, that most of your conclusions were erroneous." While Sherlock agrees with Watson that Dr. Mortimer is a country doctor who does a lot of rambling, he points out that C.C.H. actually stands for Charing Cross Hospital in London. C.C.H. was a general teaching hospital originally established behind the Haymarket Theatre in 1818, the year *Frankenstein* was published. The hospital was the dream of a Dr. Benjamin Golding, partly educated at the University of Edinburgh, who wanted to establish a place of healing for the poor. The stick was a gift from C.C.H. to Dr. Mortimer upon his leaving London to practice in the country. Given that most doctors probably wouldn't relinquish a position at such a hospital to work in the sticks, Sherlock concludes that Mortimer most likely had a humble position at C.C.H., maybe little more than a student. Sherlock concludes that Dr. Mortimer is a *young* doctor, and not the physician of more advanced age that Watson had forecast. Finally, Sherlock deduces that Mortimer was the owner of a medium-sized dog. This deductive claim was so bold and unexpected that it made Watson ejaculate with laughter, yet when Mortimer next shows up, the reader finds that Sherlock was indeed correct, and the tell-tale sign that led Sherlock to such a canine conclusion was the dog's teeth marks on Dr. Mortimer's stick.

Indeed, the history of the relationship between Sherlock and Watson is one steeped in deductive flashes of forensic brilliance. In the BBC *Sherlock* episode "A Study in Pink," Sherlock turns his prowess on John's smartphone:

> **Sherlock (holding out his hand):** "Your phone. It's expensive, email enabled, MP3 player, but you're looking for a flatshare—you wouldn't waste money on this. It's a gift, then."
>
> (At this point, John has given Sherlock the phone and he turns it over and looks at it again as he talks.)
>
> **Sherlock:** "Scratches. Not one, many over time. It's been in the same pocket as keys and coins. The man sitting next to me wouldn't treat his one luxury item like this, so it's had a previous owner. Next bit's easy. You know it already."
>
> **John:** "The engraving."
>
> (We see that engraved on the back of the phone are the words: Harry Watson, From Clara xxx.)
>
> **Sherlock:** "Harry Watson: clearly a family member who's given you his old phone. Not your father; this is a young man's gadget. Could be a cousin, but you're a war hero who can't find a place to live. Unlikely you've got an extended family, certainly not one you're close to, so brother it is. Now, Clara. Who's Clara? Three kisses says it's a romantic

attachment. The expense of the phone says wife, not girl-friend. She must have given it to him recently—this model's only six months old. Marriage in trouble then—six months on he'd just given it away. If she'd left him, he'd have kept it. People do—sentiment. But no, he wanted rid of it. He left her. He gave the phone to you: that says he wants you to stay in touch. You're looking for cheap accommodation, but you're not going to your brother for help: that says you've got problems with him. Maybe you liked his wife; maybe you don't like his drinking."

John: "How can you possibly know about the drinking?"

Sherlock (smiling): "Shot in the dark. Good one, though. Power connection: tiny little scuff marks around the edge of it. Every night he goes to plug it in to charge but his hands are shaking. You never see those marks on a sober man's phone; never see a drunk's without them."

The above analysis of John's smartphone from *Sherlock* is near-identical to a passage in the original tale where Sherlock shows off his deductive process on Watson's watch, but one important detail is altered. In the BBC rendering, Sherlock actually gets something wrong. In both versions, Sherlock guesses that Harry is John's brother. He's right in the original story, but in *Sherlock*, as this is now the twenty-first century, the BBC does not shy away from depicting LGBT+ characters textually. Harry is short for Harriet—John's lesbian sister, whom Sherlock at first mis-recognized as a straight man.

The Deductive Technique

What's the science of deduction that Sherlock uses in the field? In other words, what's the scientific context of the Deduction Diamond for these Sherlock performances in Doyle's work, and how does Sherlock reach such amazing results? As we know, Sherlock covets the curiosity of a consulting case. In Conan Doyle's second novel, *The Sign of Four*, Sherlock is lounging on a couch, as bored and as listless as one could be. He craves a new case, a conundrum to challenge his languorous mind. "My mind rebels at stagnation. Give me problems, give me work, give me the most abstruse cryptogram or the most intricate analysis, and I am in my own proper atmosphere . . . But I abhor the dull routine of existence. I crave for mental exaltation."

Watson hands his watch to Sherlock, who then conjures a deduction that both amazed Conan Doyle readers at the time and inspired the smartphone homage by the creators of *Sherlock* over a century later. In *The Sign of Four*, Watson originally thought that Sherlock's task would be an impossible ask. Watson also hoped to admonish Sherlock with a lesson "against the somewhat dogmatic tone which he occasionally assumed," but he was to be disappointed. When Sherlock finally snapped the case to and handed it back, he declared, "The watch has been recently cleaned, which robs me of my most suggestive facts. Though unsatisfactory, my research has not been entirely barren. Subject to your correction, I should judge that the watch belonged to your elder brother, who inherited it from your father." Watson replies by saying that he assumes Sherlock deduced this from "the H. W. upon the back." Sherlock confirms this fact, and goes much further:

> The W. suggests your own name. The date of the watch is nearly fifty years back, and the initials are as old as the watch: so it was made for the last generation. Jewelry usually descends to the eldest son, and he is most likely to have the same name as the father. Your father has, if I remember right, been dead many years. It has, therefore, been in the hands of your eldest brother.

Upon encouragement from Watson, Sherlock then delivers some decisive deductive blows to Watson's hopes of admonishment. "He was a man of untidy habits—very untidy and careless. He was left with good prospects, but he threw away his chances, lived for some time in poverty with occasional short intervals of prosperity, and finally, taking to drink, he died. That is all I can gather."

The story then goes on to reveal Sherlock's technique. It's the scientific meaning in terms of the Deduction Diamond. Sherlock says that he never guesses, as it's a shocking habit, and destructive to his logical senses. He suggests that his technique may appear amazing and strange to those who observe it because they neither follow his train of thought nor see the minutiae of detail upon which the strange inferences depend. Take Sherlock's statement of Watson's brother being careless, for example. Upon what minutiae did this deduction depend? The lower part of the watchcase is not only dinted in two places but is also cut and marked all over from the owner's habit of pocketing other hard objects, such as keys or coins, so Sherlock concludes that a man who treats an expensive watch in so cavalier a manner must be pretty careless.

Sherlock also describes how he came to the conclusion that Watson's brother lived for some time in poverty. He explains to Watson that pawnbrokers in England, when they take in a watch, have a habit of scratching the number ID of the pawn ticket with a pinpoint on the inside of the watch case. Sherlock says this habit works well as there's no risk of the number being lost or mislaid. He then informs Watson that there are four such numbers visible on the inside of Watson's watch case. Conclusion? That Watson's brother was often stone broke. A secondary conclusion was that Watson's brother must also have had times of relative prosperity, or else could not have redeemed the watch.

Finally, Sherlock tells Watson to take a look at the watch's inner plate, which contains a keyhole. Sherlock then suggests that the thousands of scratches around the hole are marks made when the key has slipped. It's a huge number of marks, so what sober man's key would have scored such grooves? Surely this is the handy work of a drunkard? Sherlock says one rarely sees a drunkard's watch *without* such myriad marks. Then, leaving Watson stunned, Sherlock finishes with "Where is the mystery in all this?"

How Deduction Works

Let's dig a little deeper into the scientific context of the Deduction Diamond with regard to Sherlock's methods. Sherlock's deduction technique is a fusion of deductive, inductive, and abductive reasoning. Many of his amazing conclusions are the result of a series of these logic models. Let's review these three pillars of reasoning before looking at the watch deduction in more detail.

Deductive Logic

The kind of logical argument that applies deductive reasoning to arrive at a conclusion is known as a syllogism, after the Greek word *syllogismos*, meaning "conclusion" or "inference." In its earliest form, a syllogism comes about when two propositions truly imply a conclusion. For instance, knowing all men are mortal (major proposition) and that Sherlock is a man (minor proposition), Watson may correctly conclude that Sherlock is mortal. Such syllogistic arguments can be coded in the following way:

Major proposition: Every X is Y.
Minor proposition: Z is X.
Conclusion: Thus, Z is Y.

This example can be summarized in a three-line form: All men (X) are mortal (Y). Sherlock (Z) is a man (X). Thus, Sherlock (Z) is mortal (Y).

Inductive Logic

Inductive logic is a way of reasoning in which a set of observations is synthesized to develop a general principle. Inductive reasoning differs from deductive reasoning. As we have seen, if its propositions are correct, the conclusion of deductive logic is certain. In contrast, the conclusion of inductive logic is probable, based upon the evidence given. Inductive arguments can be coded in the following way:

Specific observation: X is observed.
A second specific observation: X is observed.
A further specific observation: X is observed.
Conclusion: Based on previous observations, it is expected that X will be observed again.

Example: Sherlock pulls a cocaine throat lozenge from his pocket. (This shouldn't be a surprise. Back in the day, popular home remedies for all sorts of ailments included the likes of cocaine tablets.) Sherlock pulls another cocaine lozenge from his pocket, and yet another. Conclusion: Sherlock has a lot of cocaine lozenges in his pocket.

Abductive Reasoning

Abductive reasoning or inference begins with an observation, or set of observations, and then seeks the simplest and most likely conclusion from those observations. For Sherlock's detective work, this type of reasoning is useful for forming hypotheses to test against. Truth is, Sherlock often uses abductive reasoning in the tales of Conan Doyle, even though it's referred to as "deductive."

Take the example of *The Red-Headed League*. In that tale, Sherlock rightly identified Jabez Wilson's current and past trade, his having spent time in China, and his membership in Freemasonry. If Sherlock had used deduction for this last of the three conclusions, it would have gone something like this: (a) everyone wearing an arc-and-compass breastpin is a Freemason; (b) Jabez Wilson is wearing an arc-and-compass breastpin; (c) thus, Jabez Wilson is a

Freemason. The trouble with such an approach is this: Jabez Wilson might be wearing the breastpin *without* being a Freemason. So, the conclusion is not necessarily true, as the basis for the argument is flawed.

Using the abductive method, Sherlock's conclusion would be based on the best explanation possible with the data known, so his actual train of thought in the case of *The Red-Headed League* would be more like this: (a) the interesting fact of an arc-and-compass breastpin on Jabez Wilson is observed by Sherlock; (b) but if Jabez Wilson is a Freemason, an arc-and-compass breastpin on the man would be a matter of course; (c) thus, there is reason to suspect that Jabez Wilson is a Freemason.

The data gathering of the abductive method is most closely associated with the physical sciences and its adherence to the scientific method to increase the accuracy and reliability of both the observations and the conclusions drawn from them. One example of abductive reasoning is a jury decision. Imagine the surreal situation of Sherlock sitting on a jury. The defendant in our hypothetical case is the image of the man Sherlock just happened to see robbing the bank. Not only that, but the defendant stutters and pauses like a guilty party. The jury indeed concludes that he is guilty, but Sherlock doesn't. Sherlock was tempted to make a similar decision based on his observations but is not certain it is the right decision.

Sherlock's Watch Deduction

Having reviewed the three pillars of reasoning, lets now look at Sherlock's watch deduction in more detail, exploring some of the inferences and conclusions he reaches. A possible sequence of Sherlock's logical thoughts might run like this:

Deduction 1
Major proposition: Brand new watches appear polished and have no scratches.

Minor proposition: Watson's watch has scratches.

Conclusion: Watson's watch is not new.

Observation: The logic and argument are valid, as the propositions are true. No Victorian watchmaker worth his salt would purposely make a watch riddled with scratches.

Deduction 2

Major proposition: Folks who permit their property to be scratched are careless.

Minor proposition: Watson's watch is scratched.

Conclusion: The watch owner was careless.

Observation: This argument is valid, but the major proposition is false, and so the conclusion is false. The mere appearance of scratches on the watch doesn't necessarily mean the owner is careless. The owner could either have been in an accident and the watch dropped. Or maybe the owner bought the watch secondhand, the scratches already on it. Furthermore, considering that the watch is over fifty years old, one might expect a little wear and tear, some of which may be irreparable.

Inductive Reasoning

Sherlock's thought process: Well, *I* own a watch, and I personally make damn sure that I don't spoil or scratch the timepiece by placing it into a pocket replete with the flotsam and jetsam of life, like keys and coins. I've seen Watson look after his watch by doing the same—placing it into a pocket specifically devoid of keys and coins.

Observed pattern: Careful people who wish to keep their watches safe do so by placing them in pockets separate from the pockets in which they keep their coins and keys so that their watches aren't scratched. People who do keep their watches in pockets with keys and coins are careless.

Abductive Reasoning

London pawnbrokers engage in the routine habit of scratching the number ID of the pawn ticket with a pinpoint on the inside of the watch case. Sherlock observes that Watson's watch has four different sets of numbers. This suggests that whoever was the owner of the watch probably sold and bought it back from the brokers on at least four different occasions. This is a case of abductive reasoning coupled with Sherlock's legendary observation skills and expert knowledge of London's street life, possibly through the Baker Street Irregulars. Through them, Sherlock would know that not everyone in Victorian London had the money to own a windup watch, or thus had the bad luck to be able to sell one to a pawnbroker in the first place. Furthermore, the watch's value, along with the frequency with which the watch is sold and bought back, suggests that this timepiece not only had a monetary value but also a sentimental one.

Further Abductive Reasoning

People who are prudent with their money never have to sell and buy back their watch, especially not multiple times. As Watson's watch has been sold and bought back four times, perhaps the owner has a reasonably steady income but has debts or some financial problem which means they have a cash flow problem above their normal means.

Deduction 3

Major proposition: Sentimental people buy back pawned but cherished items.

Minor proposition: Watson's watch owner sold the timepiece only to buy it back. Conclusion: Watson's watch owner is sentimental.

Further Abductive Reasoning

It's curious, this case. We are speaking of a watch owner who owns a valuable timepiece but has to sell and rebuy it on four different occasions. Clearly, the watch has monetary value, as the pawnbroker bought it on each of the four occasions, so it must be worth selling. Yet, for the watch owner to be able to buy the watch back again suggests that money has, once more, come back into their life. If not a gambling man, nonetheless this financial situation indicates that the watch owner is perhaps not great with their money, but they must be wealthy enough to repurchase the timepiece for mere sentimental reasons.

Abductive Reasoning and Inductive Reasoning Combined

The final part of Sherlock's watch deduction centers around the drinking habits of the watch owner. Sherlock first uses abductive reasoning to figure out how the scratches may have been made. He sums up what he knows so far about the timepiece and its owner: a person who is in and out of money, and who is often careless with an expensive timepiece. Perhaps the person winding the watch has a medical condition that causes their hands to shake. Or maybe it's their age which causes them to shake. However, given that Sherlock knows that the watch was the property of Watson's older brother, he can make a confident assumption that the problem isn't medical in nature, as Watson gave no such word of a medical problem in the family. And as Watson's brother can't be much older than Watson himself, it's unlikely to be age as the root cause of the shaking. So, what would cause Watson's brother's hands to shake as he wound the watch? Since Sherlock had previously seen other scratched watches (inductive reasoning) due to the watch owner's drinking, Sherlock can come to the conclusion that Watson's brother has a drinking problem.

Now that we've used the scientific context of the Deduction Diamond to look at Sherlock's brilliantly sharp observation skills and well-tuned logical powers, let's now look at the contribution Conan Doyle's prime detective made to the field of forensic science (even before the field existed).

CHAPTER FOUR

AN ADVENTURE IN FINGERPRINTS AND FOOTPRINTS

. . . in which we begin to look at Sherlock's contribution to the field of forensic science, the art of observation, governed by science.

Many of the methods invented by Conan Doyle are today in use in the scientific laboratories. Sherlock Holmes made the study of tobacco ashes his hobby. It was a new idea, and now every laboratory has a complete set of tables giving the appearance and composition of many ashes. Mud and soil from various districts are also classified much after the manner that Holmes described. Poisons, handwriting, stains, dust, footprints, traces of wheels, the shape and position of wounds, the theory of cryptograms—all these and other excellent methods which germinated in Conan Doyle's fertile imagination are now part and parcel of every detective's scientific equipment.

—*Illustrated London News,* February 27, 1932

Who Put These Fingerprints on Sherlock's Imagination?

In Conan Doyle's tales of Sherlock Holmes, there are a number of cases in which fingerprints are commented upon, though they aren't actually used to nail a perpetrator. In *The Adventure of the Norwood Builder*, Inspector Lestrade asks "You are aware that no two thumb-marks are alike?"

In terms of the scientific context of the Deduction Diamond, what history of fingerprinting did Doyle have to depend upon for his Sherlock tales? Well, it turns out that the history of fingerprinting is a long and fascinating one. In ancient Babylon, founded around 2300 BC by the ancient Akkadian-speaking people of southern Mesopotamia, fingerprints were used on clay tablets for business transactions. The Babylonians also used fingerprints to seal contracts. Chinese records from the Qin Dynasty, between 221 and 206 BC, show details of how handprints can be used as evidence during burglary investigations. The Chinese also considered the impression of a fingerprint on a document to be a unique signature. And in fourteenth-century Persia, a book entitled *Universal History* by Khajeh Rashiduddin Fazlollah Hamadani detailed the practice of identifying persons from their fingerprints. In 1788, German anatomist Johann Christoph Andreas Mayer declared that individual fingerprints are unique. English engraver Thomas Bewick in 1804 and again in 1818 used his wood engraved fingerprints as a trademark of his work.

The Fingerprints of Jekyll and Hyde

The systematic use of fingerprints to catch criminals can be traced to Victorian England. It came in the form of a letter to the editor of *Nature* on October 28, 1880. The letter was from one Henry Faulds, born in Beith, North Ayrshire, into a family of modest means. After a difficult childhood, at the age of twenty-one Faulds enrolled at Glasgow University where he studied math, logic, and the classics. He eventually became a Scottish medical missionary at Tsukiji Hospital in Tokyo. Faulds observed that a hospital thief carelessly left a fingerprint on a wall that didn't match the print of the chief suspect. It did, however, match a secondary suspect who eventually confessed. Faulds also pointed out that monkey fingerprints were similar to their human counterparts, and that heredity may play some part in determining the form of fingerprints. Indeed, Faulds used a word to describe a fingerprint feature, the *whorl*, which is still used to this day. He concluded that fingerprints might conceivably be used to identify criminal perpetrators. He even used Jekyll and Hyde as a test case. Faulds pointed out that when Dr. Jekyll underwent his chemical-induced metamorphosis into Mr. Hyde his fingerprints would stay the same. This literary example showcased Faulds's theory that fingerprints remain unaltered throughout a person's life, the "forever-unchangeable finger-furrows," as Faulds called them.

Nature replied to Faulds on November 25, 1880. The reply came in the shape of a response from one W. J. Herschel, a British civil servant in Bengal, who claimed that he'd been taking fingerprints for over twenty years. His habit began in 1860, and his aim was to identify government pensioners trying to double-claim their pensions. Once the word got out that Herschel had begun his fingerprints project, the canny pensioners dropped their duplicity. Herschel turned his attention to jails. But Herschel differed with Faulds in some regards. Herschel held that fingerprints could *not* be used to decipher genetic relationship or ethnicity, and reported distinct differences of fingerprints within families.

Getting the Dabs to Darwin's Cousin

In the same year, 1880, Faulds wrote to Charles Darwin about his fingerprints research, and Darwin passed the news on to his cousin Francis Galton. Galton, knighted by the British Empire in 1909, was an English polymath, a tropical explorer, geographer, inventor, meteorologist, and protogeneticist. A proponent of social Darwinism, eugenics, and scientific racism, Galton was so impressed with the fingerprints research that he sought Herschel out, who gladly handed his findings over.

Galton was the Victorian apostle of quantification. He described his own early training in forensic medicine as having, "a sort of Sherlock Holmes fascination." No one scientist better personified the era's infatuation with numbers than Galton. As he was independently rich, he was able to devote his time and energy to the measurement of many things, but especially focused on trying to quantify the Victorian obsession with social class. For example, he set upon the idiosyncratic task of making a "beauty map" of Britain by classifying "the persons I meet into three classes—good, medium, and bad." He did this by a series of prick holes in paper, classifying "the girls I passed in streets or elsewhere as attractive, indifferent, or repellant." His dubious conclusion was, "I found London to rank highest for beauty; Aberdeen lowest."

Fingerprint Technology: Galton and Sherlock

By the time Conan Doyle published *The Adventure of the Norwood Builder* in 1903, fingerprinting was established as a potent forensic technology, and one that could be used in a court of law to prove a perpetrator's guilt. The year before *Norwood Builder*, a fingerprint had been accepted in an English court as evidence for a burglar's presence at the scene of the crime. By the year after *Norwood Builder*, Scotland Yard was processing up to three hundred fingerprint cards every week.

Yet, Conan Doyle was careful that Sherlock should not get carried away with too much enthusiasm for fingerprint technology. Perhaps because the fingerprint's authoritative status was beginning to be taken for granted, Conan Doyle made a point of undermining it. In *Norwood Builder*, Sherlock finds that a compelling fingerprint doesn't belong to the suspect, but is rather a forged print used to frame him.

In Sherlock's fictional world, he published on fingerprints in anthropological journals. In reality, Galton was the most prolific author of texts on fingerprinting, and he found Sherlock's methods suspect. Shortly after the *Norwood Builder* was published, Galton somewhat pedantically wrote to Conan Doyle. How, Galton asked, could the wax mold of a seal possibly have left a decipherable and bloody fingerprint on a wall? The real researchers knew that it was not possible to get "good impressions from a hard engraved material upon a hard uneven surface." Galton was clearly impressed by Conan Doyle's depiction of forensic science—up to a point. But Galton wanted further details of any experiments with fingerprints. It is not known whether Conan Doyle replied and perhaps unlikely that he ever conducted such experiments.

Sadly, this intriguing piece of correspondence came to very little, though Galton's own career in the emergent fingerprint technology was frequently compared to the fictional detective's. But Galton had huge ambitions for fingerprinting tech. To racial theorists like Galton, the fingerprint was way more than a criminal trace. For him, the goal of his unending study program into fingerprints was the expectation that each fingerprint would prove a fossil record in miniature—exuding evidence of the genealogy, character, criminality, and perhaps even the destiny of a person and peoples.

Galton and Detective Science

In 1882, a mere five years before Conan Doyle wrote the first Sherlock tale, Alphonse Bertillon, a clerk in the Prefecture of Police at Paris, created a system of classification known as anthropometry, or the Bertillon System, using measurements of parts of the body. Bertillon's system included the following: measurements of head length and head width, length of the middle finger, length of the left foot, and length of the forearm from the elbow to the tip of the middle finger. Bertillon also set up a system of photographing faces, the technique that eventually led to what we now know as mug shots. When Bertillon was made chief of the newly created Department of Judicial Identity in 1888, he used his system as the primary means of identification. This included fingerprinting, but as a secondary role in the category of special marks.

In the same year of 1888, a year after Conan Doyle published the first Sherlock tale, a *Pall Mall Gazette* journalist visited Galton's

Anthropometric Laboratory in London. Banked instruments told the tale of Galton's obsession: to measure the mental and physical traits, from sharpness of hearing to lung capacity, of every visitor to the lab. Indeed, the AnthroLab had seen over ten thousand such visitors. The *Pall Mall* report, titled *A Morning with the Anthropometric Detectives*, painted a picture of Galton's lab as a world of "order and precision, and tests of the nicest accuracy." Galton told the reporter that, though dumb, the instruments might prove to be splendid detectives. The reporter replied with much enthusiastic fear. "Splendid detectives! I am not, I hope, in a department of the Criminal Investigation Department, unconsciously yielding convincing proofs of personal identity with some scoundrel hitherto unhung."

The parallel between science and detection in the *Pall Mall* report was no accident. The technical evolution of anthropometric instruments like Galton's developed swiftly with the rise of the detective novel, and especially Conan Doyle's. Some writers were ignorant of the fact that Conan Doyle's thoughts on fingerprints were as progressed as Galton's. For example, one reporter for the *British Medical Journal*, clearly unaware of several Conan Doyle stories on fingerprints, wrote that he would be amazed if "Conan Doyle, the creator of Sherlock Holmes, does not work something up by Mr. Francis Galton's unique book on thumb-marks." In another example, *The Humanitarian* even suggested that fingerprint technology "opens up a new avenue for the Braddons and Gaboriaus, and other 'detective novelists,' to track and run to earth the most mysterious crimes. A field of investigation is here opened which will rival in fascination the celebrated 'Adventures of Sherlock Holmes.'" The reporters for both journals were correct, of course, but appear to have missed the important point that Conan Doyle was already ahead of the game.

Galton, whom we might dub the anthropometric detective, was somewhat similar to Sherlock in that he married his forensic knowledge with hypotheses and interpretations about unsolicited documents and cadavers. But, unlike Sherlock, Galton's mission was less about solving crime than the hopeful expectation that fingerprints could help uncover latent and inherited traits. Meanwhile, through Sherlock, Conan Doyle could realize the fantasy of a textual body unconsciously inscribing its traces and traits in a way that Galton never could, as Galton was tethered by the disappointing facts of empirical data.

Fingerprint Technology Goes Global, Eventually

By 1892, the year that the run of *The Adventures of Sherlock Holmes* stories was complete and in print, Galton's seminal work, *Finger Prints*, was also published. Galton tried to verify the veracity of a person's fingerprints as being unique. This work led to the setting up of a committee to look into the possibility of adopting fingerprint technology as a way of identifying criminals. The committee's classification system for fingerprints was established in 1901. It was named the Henry System, after committee member Sir Edward Richard Henry, who had used the system to such great effect and such acclaim that he was made commissioner of Scotland Yard.

Though the Henry system had been adopted throughout Europe, it was not all a story of success. In 1909, a man named Oscar Slater was wrongly imprisoned for eighteen years for the murder of Marion Gilchrist despite the fact that a bloody handprint had been left on a chair at the murder scene. The case is notable for two reasons. One, it's clear that Scotland Yard were still not completely using fingerprinting. Two, Conan Doyle had become personally involved in the case to prove Slater's innocence, publishing *The Case of Oscar Slater* in 1912.

The main problem with fingerprint technology identification in the early days was, of course, the almost impossible task of finding a fingerprint match from manual searches. Even when a significant number of criminals had been fingerprinted, just think about the amount of time taken to search through the millions of prints in manual files. The time needed to locate the correct fingerprint was gargantuan. What's worse, many of the early recorded prints were unsystematic, and of inferior quality. This situation persisted until way into the twentieth century.

It wasn't until the early 1980s that the first computer databases of fingerprints were developed. Following research initiatives across Japan, the UK, France, and the US, these systems became known as Automated Fingerprint Identification Systems (AFIS). A lot of kudos for this development goes to American police inspector Ken Moses. Moses worked out of San Francisco and was outraged in 1978 by the homicide of a forty-seven-year-old woman who had survived Nazi concentration camps only to be murdered in San Francisco. Moses had the unenviable mission of matching prints at the scene of the crime with the prints on file of over four hundred thousand people taken in the city over a period of forty-five years. Moses began his search in 1978 but, six years later, and without having moved any further along with the murder hunt, he read about computer-automated fingerprint ID systems and raised enough money to establish AFIS in San Francisco. The result? Moses got a match within sixty seconds. A couple of days later, the killer was arrested for first-degree murder.

How Sherlock Used Fingerprints

The tales of Sherlock Holmes are some of the very few places where the emergent fingerprint technology is referenced in fiction. A murderer is identified using fingerprints in Mark Twain's 1883 book *Life on the Mississippi*, but Conan Doyle wrote many Sherlock cases where fingerprints are cited. They include *The Sign of Four* (1890), *The Man with the Twisted Lip* (1891), *The Adventure of the Cardboard Box* (1893), and much later *The Adventure of the Three Gables* (1926).

In *The Sign of Four,* Sherlock points out that a thumbprint was left on the envelope sent to Mary Morstan by Thaddeus Sholto, the character in the tale who was written as an homage to Oscar Wilde. Sherlock suspects that the thumbmark was left by the postman (Sherlock: "Postmark, London, S. W. Date, July 7. Hum! Man's thumb-mark on corner,—probably postman. Best quality paper. Envelopes at sixpence a packet. Particular man in his stationery. No address.") Nothing comes of the print, as Sholto identifies himself to Mary Morstan.

In *The Man with the Twisted Lip*, another thumbprint is found. This time, it's a greasy mark on an envelope (Sherlock: "Hum! Posted to-day in Gravesend by a man with a dirty thumb. Ha! And the flap has been gummed, if I am not very much in error, by a person who had been chewing tobacco,") which holds the note from Neville St. Clair to his missus. Once more, the thumbmark proves to be of little use, as, again, it belongs to a third party; this time to an acquaintance of St. Clair's who posted the note. Sherlock cracks the case with other evidence.

Thumbprints are in the frame once more in *The Adventure of the Cardboard Box.* Actually, this time it's *two* distinguishing thumbmarks on the eponymous cardboard box, which Susan Cushing receives from Jim Browner. Once more, the prints are not used, as Sherlock cracks the case using other methods.

Finally, in *The Adventure of the Three Gables,* the police inspector on the case retains a page from Maberly's novel, as it may have prints daubed on it. In all of the above cases, either the police or Sherlock himself look for or gather fingerprint evidence as part and parcel of the detective process even though no conclusive evidentiary prints are found.

Intriguingly, there are a couple of tales where the suspicious lack of fingerprints is germane to the case. In *The Adventure of the Three Students*, Sherlock points out that there is an absence of prints on Hilton Soames's exam papers. In *The Adventure of the*

Red Circle, Sherlock suspects that the reason the instructions sent to Mrs. Warren are missing a corner is that the sender was tearing off a print mark:

> **Sherlock:** "You will observe that the paper is torn away at the side here after the printing was done, so that the 's' of 'soap' is partly gone. Suggestive, Watson, is it not?"
> **Watson:** "Of caution?"
> **Sherlock:** "Exactly. There was evidently some mark, some thumbprint, something which might give a clue to the person's identity."

This brings us to the 1903 Conan Doyle tale *The Adventure of the Norwood Builder* in which Lestrade declares that no two thumb-marks are alike. The context is that Lestrade has found on a wall the bloody thumbprint (yet another thumbmark) of the chief suspect John Hector McFarlane. Lestrade rather triumphantly reminds Sherlock that fingerprints are said to be unique, and so his finding of the thumbprint is hopefully instrumental to the case. Sherlock knows of this uniqueness, obviously, but is also aware that the print had been planted there after McFarlane had been taken into custody:

> **Sherlock, sarcastically:** "What a providential thing that this young man should press his right thumb against the wall in taking his hat from the peg! Such a very natural action, too, if you come to think of it."
> **Watson, as narrator:** Holmes was outwardly calm, but his whole body gave a wriggle of suppressed excitement as he spoke.

The day before, it had been Sherlock alone who carried out a forensic study of the wall concerned. Sherlock deduced that the print must have been planted on the wall under the cover of darkness by one Jonas Oldacre. His aim, to place McFarlane in the frame. Oldacre had procured McFarlane's print by asking him to press down a wax seal on a legal document. It's clear that Conan Doyle gifted the devious Oldacre with some technical intelligence, as Oldacre must have been aware of the fact that, as Lestrade says, no two thumbmarks are alike. If not, Oldacre would have made the error of planting anyone's thumbprint, maybe even his own, and not gone to the trouble of procuring McFarlane's in wax.

Here's a measure of how up-to-date Conan Doyle's forensic research was. *The Adventure of the Norwood Builder* was published in October 1903. On June 27 of that same year, an article under the heading *Criminals Convict Themselves* appeared in *Tit-Bits* magazine. The article reported on a recent case in the county of Yorkshire in England where a burglar, clearly with relaxation on his mind, had taken some time out from his apparently less-than-pressing crime and left a mucky thumbprint on a book. Did Conan Doyle read the article? It's more than likely. *Tit-Bits* was the more popular sister magazine of *The Strand*, which played a significant role in establishing Sherlock as a literary and cultural icon. Not only that, but between 1892 and 1918, *Tit-Bits* published six short stories, one novel, one interview, and three articles written by Conan Doyle himself.

Speculation aside, *The Adventure of the Norwood Builder* is the first occasion in fiction where a false fingerprint is used. The series of stories in which Sherlock references fingerprints is testament to the rapacious research that kept Conan Doyle's tales so scientifically vital. By the time London's Scotland Yard rubber-stamped fingerprint technology for criminal procedures, Conan Doyle had already penned three tales where this method was used.

A Journey Begins with a Single Footprint

Alfred Hitchcock once said, "Blondes make the best victims. They're like virgin snow that shows up the bloody footprints." While Conan

Doyle's tales may lack platinum blondes, they certainly have their fair share of footprints. From the get-go, with 1887's *A Study in Scarlet*, Sherlock was using footprints in his consulting work.

Thinking again about the scientific context of the Deduction Diamond, we find that, as with fingerprints, the use of footprints runs back to antiquity. There's an interesting legend of humans and idols in the book of Daniel, and the story is an ancestor of the "locked-room mystery." The idol concerned is Ba'al, which means "lord" in the Northwest Semitic languages spoken in the Levant during antiquity. As alleged vanquisher of the sea, Ba'al was worshiped by the Canaanites and Phoenicians as the patron of sailors and seagoing merchants. The story goes that the King of Persia had been persuaded by priests to a policy of leaving out large amounts of food at night so that Ba'al could feed.

Daniel is keen to ridicule this worship of idols who are meant to be eating so much food despite being "made of clay covered by bronze," but the priests still persuade the King to set out Ba'al's usual feast: "twelve great measures of fine flour, forty sheep, and six vessels of wine." Those idols sure can stuff themselves. However, the entrance to the temple was sealed so that if it was found that Ba'al wasn't consuming the feast, the priests would be sentenced to death. It is Daniel, the Sherlock of this story, who uncovers the ruse of the priests. He scatters ashes over the floor of the temple in the presence of the King after the priests have left. The following morning, the King inspects Daniel's ashes test by observing from above. He sees that the feast has been consumed and observes that the wax seal is still intact. But, before the King offers a hosanna to Ba'al, Daniel shows him the footprints on the temple floor. A set of *human* footprints, some quite slender and small as if made by children, showed not only that it had been the *priests* that consumed the feast, but that their wives and children had helped them finish off the food. The priests of Ba'al confess their deed, and reveal the secret passage they used to steal inside the temple, thereby qualifying the tale as not only a locked-room mystery but also a foremost example of the relatively forensic use of footprints.

Since those early days of myth and legend, the use of footprints has been instrumental in the forensic evidence of prehistoric life. For example, an international group of scientists discovered a series of 1.5-million-year-old human ancestor footprints in Kenya. The footprints are the earliest direct evidence of upright walking by humans. The scientists think that the prints were probably made by the species *Homo erectus*.

As with Sherlock's consulting work in crime, the scientific use of "trackways" is also fundamental to working out what happened in the past. A trackway is a set of prints left by a life-form, and could be the footprints, hoof-prints, or paw-prints of an animal. Lark Quarry in Australia boasts the world's only discovered trackway to date, which is a record of a dinosaur stampede that happened around one hundred million years ago. A set of footprints in the Chania Prefecture in Crete is evidence of hominin-like footprints from the Miocene era, some 5.7 million years ago. Finally, way before the days of Daniel and his ashes trick, human footprints from 3.7 million years ago were preserved in volcanic ash in Laetoli in Tanzania.

In paleontology, a series of two or more footprints can be informative. They can reveal whether an animal walked on two legs (bipedal) or four legs (quadrupedal). They can also indicate whether a dinosaur, for example, was walking, running, or even wading through water at the time it laid the tracks. By measuring details like the length of a single step (a pace) and the distance between placements of the same foot (a stride), the dinosaur's approximate hip height, size, and speed can be determined using formulas. By combining the insights with other details of the ground and surroundings, scientists can determine what the animals appear to be doing and what other things were in their environment at the time.

Scientists do something similar when reading forensic footprints. Prints from footwear can be left on most surfaces, from paper to the human body itself. Prints are separated into three kinds: visible, plastic, and latent. A visible print is the kind Sherlock would have seen. They occur in those cases where there is a transfer of material from the footwear directly to the surface. This kind can be seen by the naked eye without additional tech, so would include bloody shoe prints left in a murder room. A plastic print is a 3D impression left on a soft surface. This includes shoe tracks left in sand, mud, or snow, or like the boot-print of Neil Armstrong on the lunar surface. Finally, a *latent* print is the kind that's not visible to the naked eye. Latent prints are made by static charges between the sole and the surface. Forensic technicians use chemicals, powders, or different light sources to locate these prints. Examples of latent prints are those shoe prints detected on tile or hardwood floors, windowsills, or metal counters.

How Sherlock Used Footprints

While Sherlock references fingerprinting in only seven of Conan Doyle's sixty stories, the use of *footprints* crops up in twenty-six of them. Understandably, given that fingerprint technology was still in development and footprint forensics can more easily be realized in the minds of the readers. We spoke earlier about Conan Doyle's battle to keep the Sherlock stories fresh and avoid the tales reading as a tad formulaic. To this end, the footprint, as an item of case evidence for Sherlock, is found in a range of different materials.

The footprint is found in clay soil in *A Study in Scarlet*, in a carpet in *The Adventure of the Resident Patient*, in blood in *The Adventure of the Red Circle*, in snow in *The Adventure of the Beryl Coronet*, in mud and in dust in *The Sign of Four*, on a curtain in *The Adventure of the Crooked Man*, and in ashes in *The Adventure of the Golden Pince-Nez*. In all cases which include footprints, Sherlock uses them as evidence, with varying degrees of success.

Sherlock's most convincing use of footprints includes his first two novels, *A Study in Scarlet* and *The Sign of Four*. Here, Sherlock's footprint acumen is so successful in tracking people that those concerned are unsettled by his uncanny use of technique. In *A Study in Scarlet*, Sherlock is able to accurately track the movements of the constable John Rance. At one point, Watson declares:

> There were many marks of footsteps upon the wet clayey soil, but since the police had been coming and going over it, I was unable to see how my companion could hope to learn anything from it. Still I had had such extraordinary evidence of the quickness of his perceptive faculties, that I had no doubt that he could see a great deal which was hidden from me.

Sherlock explains his technique later in the tale:

> No doubt it appeared to you to be a mere trampled line of slush, but to my trained eyes every mark upon its surface had a meaning. There is no branch of detective science which is so important and so much neglected as the art of tracing footsteps. Happily, I have always laid great stress upon it, and much practice has made it second nature to me. I saw the heavy footmarks of the constables, but I saw also the track of the two men who had first passed through the garden. It was easy to tell that they had been before the others, because in places their marks had been entirely obliterated by the others coming upon the top of them. In this way my second link was formed, which told me that the nocturnal visitors were two in number, one remarkable for his height (as I calculated from the length of his stride), and the other fashionably dressed, to judge from the small and elegant impression left by his boots.

Sherlock does the same for the story's main antagonist, Jonathan Small, and his diminutive and loyal companion Tonga in *The Sign of Four*. Small is so shocked by Sherlock's detection that he says to Sherlock, "You seem to know as much about it as if you were there!"

Naturally, in this particular case, Sherlock's efforts are made easier by the fact that Small has a peg leg, part crippled by a hungry crocodile biting off his leg, and that Tonga is a pygmy indigenous to the Andaman Islands. When Sherlock first unveils Tonga's elfin footprint to his sidekick, Watson declares horror that a child should have "done this horrid thing!" Sherlock knows better, but perhaps Watson can be forgiven to jumping to such a conclusion in a megalopolis like London when it had been reported that the city could boast tens of thousands of deserted, roaming, and lawless children.

The question of distinctive tracks pops up again in *The Adventure of the Beryl Coronet*. This tale has a foursome of footprints. First, there's the boot print belonging to the dissolute playboy and gambler Sir George Burnwell. Next, there's the shoe print of servant girl Lucy Parr, and the naked footprint of Arthur Holder, young friend of Burnwell's. Finally, there's the wooden leg of Francis Prosper, green-grocer sweetheart of Lucy Parr's. After suggesting that "a very long and complex story was written in the snow" in front of him, Sherlock proceeds to use the footprint profile to divine the movements in the snow. First, Sherlock concludes that Lucy Parr and Arthur Holder ran. Folks that are running leave tracks which are usually deeper at the front than at the back. Second, he says that Francis Prosper had been meeting with Lucy Parr, who ran when discovered. None of the three above had anything to do with the thieving of the beryl coronet. That responsibility was down to Sir George Burnwell. He had looted the crown and been quickly chased by Arthur Holder. Indeed, this forensic reconstruction of the crime scene, based on Sherlock's footprint analysis, is the main evidence for reversing the verdict of Arthur Holder's culpability from guilty to innocent.

Conan Doyle again uses the idea of footprints made while running in the opening pages of *The Hound of the Baskervilles*. Along with the paw prints of the huge hound, human footprints belonging to Sir Charles Baskerville are detected along the yew alley. Sherlock worked out that Sir Charles's footprints showed he had not been on tiptoe, as has been suggested, but had been running from the hound:

> **Sherlock:** "That change in the footprints . . . What do you make of that?"
> **Watson:** "Mortimer said that the man had walked on tiptoe down that portion of the alley."
> **Sherlock:** "He only repeated what some fool had said at the inquest. Why should a man walk on tiptoe down the alley?"
> **Watson:** "What then?"
> **Sherlock:** "He was running, Watson—running desperately, running for his life, running until he burst his heart and fell dead upon his face."

Incidentally, the British Medical Journal pointed out that Sir Charles Baskerville's demise by heart attack while fleeing the infamous hound has given rise to the medical term the "Baskerville Effect," which refers to heart attacks brought on by extreme stress. Conan Doyle first described such an unenviable death in *The Sign of Four*.

In *The Adventure of the Resident Patient*, Sherlock stuns Watson by deducing, from footprints on the carpeted stairs, the order in which the guilty parties climbed the stairs. This footprint prediction comes before Sherlock accounts for their movements when they hang Mr. Bessington in his room. Meanwhile, in *The Adventure of the Devil's Foot*, Sherlock again manages to decipher two footprints with distinct characteristics. One footprint is the relatively unremarkable print of Mortimer Tregennis. The other is the ribbed tennis sneaker track of Dr. Leon Sterndale. These prints are integral to the evidence Sherlock orchestrates to show that Mortimer murdered his sister Brenda Tregennis. In 1846, French novelist Eugène Sue wrote "revenge is *very good eaten cold*, as the vulgar say," and revenge is what Dr. Sterndale exacts upon Mortimer Tregennis for causing the death of his untold love. *The Adventure of the Devil's Foot* is also one of those tales in which Sherlock cracks the case, yet lets the culprit walk away, deciding Dr. Sterndale's actions to be just.

There are a couple of cases in which Sherlock uses some technique to tease out a footprint. In *The Adventure of the Devil's Foot*, he deliberately spills some water so that he can capture a shoe print of Mortimer Tregennis. In *The Adventure of the Golden Pince-Nez*, maybe taking the book of Daniel as his inspiration, Sherlock works with tobacco ashes on a carpet to show the tracks of Anna Coram:

> **Sherlock:** "I could see no marks to guide me, but the carpet was of a dun color, which lends itself very well to examination. I therefore smoked a great number of those excellent cigarettes, and I dropped the ash all over the space in front of the suspected bookcase. It was a simple trick, but exceedingly effective."

Sherlock's furious smoking makes him appear anxious, which embarrasses Watson, who hasn't understood what Sherlock is trying to pull off.

> **Watson:** "Holmes was pacing up and down one side of the room whilst the old Professor was talking. I observed that he was smoking with extraordinary rapidity. It was evident that he shared our host's liking for the fresh Alexandrian cigarettes."

When Sherlock returns to Professor Coram's room, he observes Anna Coram's prints in the ambushing ashes.

> **Sherlock:** "We then ascended to the room again, when, by upsetting the cigarette-box, I obtained a very excellent view of the floor, and was able to see quite clearly, from the traces upon the cigarette ash, that the prisoner had, in our absence, come out from her retreat."

Indeed, the very *absence* of Anna Coram's prints on the path outside is also part of Sherlock's case. The lack of prints led him to conclude that she is hidden behind the hinged bookcase in the professor's room. An absence of prints is an ingredient in cracking other cases, too, including *The Adventure of Black Peter*, *The Adventure of the Reigate Squire*, *The Adventure of the Naval Treaty*, *The Five Orange Pips*, and *The Adventure of the Three Students*.

Meanwhile, *The Boscombe Valley Mystery* is a case that Sherlock cracks almost solely (pun intended) using footprints. This tale can be summarized as the murder of the blackmailing tenant Charles McCarthy by the land-owning John Turner. At first, suspicion centers around McCarthy's son, James, but McCarthy's maid had helpfully given Sherlock boots belonging to the McCarthy men. On analyzing the boots, Sherlock presently heads for Boscombe Pool, the scene of the crime, where he has to unearth a solution from the mass of impertinent footprints including those of Lestrade:

> **Sherlock:** "That left foot of yours [Lestrade] with its inward twist is all over the place. A mole could trace it, and there it vanishes among the reeds. Oh, how simple it would all have been had I been here before they came like a herd of buffalo and wallowed all over it. Here is where the party with the lodge-keeper came, and they have covered all tracks for six or eight feet round the body."

But immediately upon scalding Lestrade and others, Sherlock continues with his analysis of the prints:

> But here are three separate tracks of the same feet. These are young McCarthy's feet. Twice he was walking, and once he ran swiftly, so that the soles are deeply marked and the heels hardly visible. That bears out his story. He ran when he saw his father on the ground. Then here are the father's feet as he paced up and down.

Swift upon the heels of his analysis, Sherlock detects evidence of the murderer: "a tall man, left-handed, limps with the right leg, wears thick-soled shooting-boots and a grey cloak, smokes Indian cigars, uses a cigar-holder, and carries a blunt pen-knife in his pocket. There are several other indications, but these may be enough to aid us in our search.")

Despite all this data, Inspector Lestrade doesn't apprehend John Turner, so Sherlock decides, after gifting Lestrade the clinching evidence for cracking the case, to let the terminally ill Turner go free. Famous Harvard-educated Holmes scholar Robert Keith Leavitt, longtime historian of the original *The Baker Street Irregulars*, who were devoted to all things Sherlockian, has written about Sherlock's leniency with some criminals: "In the sixty cases in the Writings, there are thirty-seven definite felonies where the criminal was known to Mr. Sherlock Holmes. In no less than fourteen of these cases did the celebrated detective take the law into his own hands and free the guilty person."

Indeed, another one of those cases where Sherlock lets the perpetrator go free, and also involves a set of footprints, is *The Adventure of Charles Augustus Milverton*. Milverton is "the worst man in London," a scoundrel and blackmailer of society women. Lestrade finds prints outside Milverton's residence and is unaware that the footprints belong to Sherlock and Watson who were inside and saw that it was Lady Eva Brackwell who murdered Milverton, but nonetheless kept quiet.

How *Sherlock* Used Footprints

The forensic use of footprints is brought up to date in the BBC's *Sherlock*. In forensic science, Locard's principle says that the criminal will bring something into the crime scene and leave with something from it; both can be used as evidence. Born ten years before Conan Doyle's first Sherlock story, French criminologist Edmond Locard was a pioneer in forensic science who became known as the Sherlock Holmes of Lyon. He stated the fundamental principle of forensic science: "Every contact leaves a trace." The principle is generally understood as "with contact between two items, there will be an exchange." Criminalist Paul L. Kirk explained the principle with great panache:

> Wherever he steps, whatever he touches, whatever he leaves, even unconsciously, will serve as a silent witness against him. Not only his fingerprints or his footprints, but his hair, the fibers from his clothes, the glass he breaks, the tool mark he leaves, the paint he scratches, the blood or semen he deposits or collects. All of these and more, bear mute witness against him. This is evidence that does not forget. It is not confused by the excitement of the moment. It is not absent because human witnesses are. It is factual evidence. Physical evidence cannot be wrong, it cannot perjure itself, it cannot be wholly absent. Only human failure to find it, study and understand it, can diminish its value.

Where footprints are concerned, a typical transfer of material occurs between footwear and the ground beneath. In the BBC's *Sherlock* episode "The Reichenbach Fall," Sherlock and John investigate the kidnapping of the children of the British Ambassador to the US. Sherlock uses his knowledge of Locard's work to obtain soil samples from the shoes which caused the prints at the scene of the kidnapping. By collecting the residue from a shoe print, Sherlock uses his skills in chemical analysis to identify an array of contact materials:

> **Sherlock:** "All the chemical traces on his shoe have been preserved. The sole of the shoe is like a passport. If we're lucky we can see everything that he's been up to."
> **Lestrade:** "Chalk, asphalt, brick dust, vegetation . . . What the hell is this? Chocolate?" Sherlock: "I think we're looking for a disused sweet factory."

Sherlock's chemical analysis found signatures of not only the chalk, asphalt, brick dust, and vegetation, but also polyglycerol polyricinoleate (E476), an emulsifier made from the naturally occurring alcohol, glycerol, and fatty acids (usually from castor bean, but also from soybean oil). E476 is an ingredient found in the mercury-tainted chocolate with which Moriarty was poisoning the children. What's more, Sherlock also uses the footprints to indicate the kidnapper's height, gait, and shoe size, and uses his mind palace to find a London building site which fits the facts of brick dust and the rest of the clues.

It is testimony to Conan Doyle's scientific imagination that Sherlock was shown to cleverly crack crimes over the intervening decades using only techniques such as fingerprints and footprints, and that the BBC can riff on the same theme almost a century later.

CHAPTER FIVE

AN ADVENTURE IN ALCHEMY AND A CHEMILUMINESCENT HOUND

. . . in which we consider Sherlock's talents in the dark arts of chemistry and celebrate his resurrection.

"Chemistry is necessarily an experimental science: its conclusions are drawn from data, and its principles supported by evidence from facts."
—Michael Faraday, *Chemical Manipulation* (1842)

You are here to learn the subtle science and exact art of potion-making . . . the beauty of the softly simmering cauldron with its shimmering fumes, the delicate power of liquids that creep through human veins, bewitching the mind, ensnaring the senses. I can teach you how to bottle fame, brew glory, even stopper death.
—Severus Snape, *Harry Potter and the Sorcerer's Stone* (1997)

Sherlock's Age of Chemistry

Sherlock's Victorian era was one in which great strides were made in chemistry. One of the youngest of the natural sciences, chemistry's growth and evolution took place almost entirely in the 1800s, so it's hardly surprising in terms of our Deduction Diamond, given the factual scientific context within which Doyle's texts were made, that he made chemistry Sherlock's first scientific love. Watson announces in *The Adventure of the Three Students* that without his chemicals Sherlock is "an uncomfortable man." Through the eyes of Sherlock, set at the turn of the nineteenth century into the twentieth, let's take a magical mystery tour of chemistry, a potted history up to that point.

There had been no such science as chemistry in the classical age of the ancient Greeks and Romans. The Greeks believed in only four elements: earth, air, fire, and water. Think about a stone falling into a pool. The Greeks thought that the stone plunges down to the pool's depths because its constituent, earth, is the heaviest of the four elements and "down" is its natural place to be. The bubbles that appear stuck to the stone on its descent are made of air. As the stone plummets, the bubbles draw up through the water to their own element of air, as "up" is air's natural place. If a fire was lit beside that pool, the flames from the fire appear to go naturally upward.

Greek Elements

The ancient Greek idea was that the four elements each had their own natural resting place. Now, as ideas go, the Greeks made quite a big blunder, though admittedly a forgivable one, as they were living over two thousand years ago. Yet the idea held sway for two millennia because their conception of an element still stands in Sherlock's day: an element is an entity to which other compounds can be resolved, but which cannot itself be resolved into anything simpler. Elements are "uncuttable."

The Greeks gave us another advantage in their basic elemental theory. Earth, water, and air also represent the three states of matter: solid, liquid, and gas. They're observable qualities that can be easily seen by everyone (and Sherlock is very fond of observable qualities). What's more is that the application of heat appears to provide evidence that these elements flow from one into the other. Ancient Greeks were, of course, aware of solids such as gold, silver, and copper, but they believed the transmutation of such elements could be achieved by simply adjusting the amounts of fire or earth or water that they contained. Such ideas led to alchemy.

Alchemy to Chemistry

The roots of Sherlock's beloved chemistry were to be found in alchemy, going way back, before the modern days of the ideas of atoms and elements, even if it wasn't always called by the name of chemistry. In *The Adventure of the Gloria Scott*, one of only two tales which Sherlock himself narrated, Sherlock says, "during the first month of the long vacation, I went up to my London rooms where I spent seven weeks working out a few experiments in organic chemistry." He would have been well aware of the fact that, in trying to transmute one classical element into another, early alchemists experimented with materials, and that these experiments helped evolve chemical techniques. The process of distillation was developed to turn a liquid into a gas, then making it liquid again by cooling. The fashion of fractionation helped separate constituent quantities from a mixture, and the method of crystallization helped turn a liquid into a crystalline solid.

Like other human endeavors, science is a living body of knowledge. While great works of literature and art live on through arbitrary interpretation, the works of science can be verified by direct and repeatable experiments in the material world. Whether present or past, each advance in science can be exposed to experiment and examination. The success of science lies in its ability to apply its schema to inanimate material systems in the case of chemistry, or living organisms in the case of biology.

It's interesting and instructive to the detective's mind to look at the way in which this living body of knowledge has evolved, to examine the pattern of scientific advance. History shows a distinct sequence of the emergence of the different disciplines of science. Generally speaking, the order is: mathematics, astronomy, mechanics, physics, chemistry, and biology.

This developmental time sequence of the sciences appears to fit very well to the patterns of social advance. Notice how the sequence corresponds quite closely to the practical uses that were expected, if not demanded, of science by ruling classes at different times. In ancient times, science derived from the techniques that arose from humanity's concern with the natural environment. For example, from the beginning of recorded history and the development of surplus, mathematics arose out of the need to make calculations relating to taxation and commerce or to measure land. Observations of the sky were used to determine the seasons, an important factor in knowing when to plant crops, as well as in understanding the length of the year. These priestly functions gave rise to astronomy, of course. Only much later did interest arise in the control of inanimate forces. The demands of the new textile industry and emerging manufacturers of the eighteenth century gave rise to contemporary chemistry.

Even after the scientific revolution, difficulties of chemical identity still occurred. Chemists identified some elements which later turned out to be compounds, and the whole painstaking process was protracted. The revolution associated with Galileo and Newton helped. Though theirs was a revolution in physics, they also helped usher in a revolution in measurement. But, for burgeoning chemists like Sherlock, challenges remained such as measuring the nuance of chemical composition and the vagaries of color, smell, and other characteristics that chemical materials might possess.

Sherlock's London had been home to a towering figure in the experimental art of chemistry in the 1660s. Robert Boyle, Anglo-Irish natural philosopher, is best known for Boyle's Law and was the seventh son and fourteenth child of the fabulously wealthy and landed Richard Boyle, 1st Earl of Cork. But Robert Boyle is also associated with the transformation from alchemy to chemistry.

The Sceptical Chymist and the French

Boyle wrote *The Sceptical Chymist* in 1661. Published in the form of a dialogue, the book presented Boyle's atomist ideas. He questioned the limited belief in just the four classical elements of the Greeks. Boyle's hypothesis was that matter was made of atoms or clusters of atoms in motion, and that every phenomenon in the cosmos was the result of collisions of these atoms.

For many decades before Sherlock, Boyle's book had become one of the most widely cited works, and it encouraged artisans and tradesmen to engage in experimental chemistry. *The Sceptical Chymist* provided a vision of the kind of projects chemists like Sherlock might achieve, as well as inventing the very experimental method that modern science has used ever since. As a result of the likes of Boyle, Britain arrived early in the game of chemistry, so Sherlock's kingdom became the country that secured the largest number of elemental discoveries (nineteen), including magnesium (in the year 1755), hydrogen (in 1766), nitrogen (1772), titanium (1791), and a whole slew of other elements in the 1800s, including almost all of the so-called noble gases.

Across the channel in France, nobleman Antoine Lavoisier had been a huge influence on the chemical revolution. Sherlock knew that Lavoisier was considered by many to be the very father of modern chemistry, and his considerable contribution to chemistry mostly stemmed from Lavoisier transforming the discipline from a qualitative to a quantitative one. In the middle of the eighteenth century, an attempt was made to explain fire, that most striking of all ordinary chemical changes. It was noted that, during combustion, there appeared to be two classes of bodies, those that burn and those that don't. The former was thought to contain the element of fire, or *phlogiston*. During burning, it was assumed that phlogiston was released into the air; the ashes of combustion left behind. During this time, the act of burning was considered to be decomposition. Combustible bodies were compound in nature, made up of phlogiston and the products of combustion. So, in the act of burning, these two elements separated, the phlogiston given off into the air, with the products of combustion left behind as the ashes.

This first theory of chemistry was replaced by a better one in the year 1785 by Lavoisier. Lavoisier found that when bodies burned, the products of combustion were actually heavier than the original substances. How can that be? Lavoisier saw that burning was the union of oxygen with the burning substance and that combustion was a chemical combination and not a decomposition. "There is no such thing as phlogiston, the element of fire," declared Lavoisier. Thus, he became famous, and understandably dubbed "The Founder of Modern Chemistry," noted for his major discovery of the role oxygen plays in combustion, identifying and naming oxygen in 1778 and hydrogen in 1783, the two combustible constituents which make up something as basic to life as water.

And so began a new era for chemistry, a quantitative era. From Lavoisier on, all substances that couldn't be resolved into simpler substances, weighing less than their originals, were labeled elements. Lavoisier also compiled the first extensive list of elements. He helped reform chemical nomenclature, so chemists about the globe like Sherlock began to call known elements by the same standardized names. He predicted the existence of silicon in 1787, and in 1777 was the first to establish that sulfur was an element and not a compound. Interestingly, at the height of the French Revolution, Lavoisier was charged with tax fraud and selling tainted tobacco and was guillotined in 1794. If only Sherlock had been there to help.

The Periodic Table

In Sherlock's century, chemists began to discover certain patterns among the elements. Chief among the most crucial developments was that of Russian chemist and inventor, Dmitri Mendeleev. Mendeleev's artful realization was not merely to spot patterns among the chemical elements and their behavioral traits. He was also able to identify where there might be gaps, which may belong to as yet unidentified elements.

Science is, to some extent, about predictions based on predictable facts. Or, in the words of Stephen Hawking, "A good theory is characterized by the fact that it makes a number of predictions." Dmitri Mendeleev's mind was prescient enough to forecast that the gaps among the elements would be filled in the future. The order in which chemical elements pop up on the table is a function of their atomic weight. Mendeleev foretold what new elements would be discovered and forecast their properties and weight. In this way, he predicted the properties of an element that would resemble boron, another that would be analogous to aluminum, and a third that would be closely related to silicon. These predictions were all realized during Sherlock's life.

Mendeleev also possessed the Sherlockian trait of arrogant self-confidence. He went so far as to suggest that elements not fitting his schema must have had their weights wrongly measured by other chemists! Indeed, he had been working so intensely on his prototype periodic table that he claimed to have conjured up a vision of the table in a dream. "I saw in a dream a table where all elements fell into place as required. Awakening, I immediately wrote it down on a piece of paper, only in one place did a correction later seem necessary," as quoted in the publication *Soviet Psychology* in 1967.

Sherlock's Victorian era was one which witnessed great advances in chemistry. Twenty-two elements were discovered in the century and a half between 1650 and 1799. The nineteenth century managed twenty-five between 1800 and 1849 alone, and a further twenty-four between 1850 and 1899, with a total of nineteen of those new elements discovered in Sherlock's homeland.

Sherlock's Organic Chemistry

Among the most startling scientific advances made in the nine-teenth century came in the branch of science known as organic chemistry. At the turn of the nineteenth century, very little was known of the compounds occurring in the organs of plants and animals. Sure, some organic substances had been separated, yet little had been gleaned of their composition, as the methods of analysis were basic at that time. Furthermore, it was previously thought that organic compounds couldn't possibly be made arti-ficially in the lab, as was the case with mineral compounds. The reason for this was partly down to *vitalism*, the belief that living things are fundamentally different from nonliving entities because they contain some ghostly mainspring, some vital, nonphysical essence, and are governed by different principles than inanimate things. Vitalism went further and suggested that this peculiar vital force somehow intervened in the production of the organs of plants and animals, and that mere humans could never expect to replicate their production in the laboratory.

The ideas of *vitalism* were eventually abandoned, and the first synthesis of an organic substance was achieved. Year after year an ever-larger number of substances was added to the list of synthetic organic compounds. It would take far too long to list all the carbon compounds that have been made artificially in the lab. Carbon today is not only known to form the basis of life on Earth, but it's also found out in deep space as a result of stellar fusion (the creation of chemical elements by nuclear fusion reactions within stars). Carbon is the backbone of biology on Earth because of its very nature. It easily bonds with life's other main elements like hydrogen, oxygen, and nitrogen. Carbon is also light and small, making it an ideal element for making the longer and more complex chemicals of life such as proteins and DNA.

None of this was known in Sherlock's day; it was yet to be discovered. Nor did they know much about carbon's proclivities.

Carbon makes one of the *softest* known substances in graphite, and also one of the *hardest* known substances in diamond. In all, carbon is known to form over ten million different chemicals, which is around 70 percent of all chemicals on Earth. In particular, carbon is made in the cores of giant and supergiant stars, about which Sherlock cares not a jot! Carbon is then scattered into space as dust in supernova explosions. Some stars use carbon as a catalyst for their fusion reactions as they burn. A number of complex carbon chemicals, including sugar, have been found in space.

Dye Me a River

The foundations of organic chemistry were still being laid. Chemists had managed to synthesize the hydrocarbons of petroleum, common alcohol (today known as ethanol), wood alcohol (today known as methanol), fusel oil (mixtures of several higher alcohols), the essential oils, the fatty acids, glycerin, some sugars, and, importantly, coloring matters and dye stuffs.

In Sherlock's day, and for a bit before, the city streets were lit by gas lamps. The gas that glowed in London's lamps was a by-product of the distillation of coal. As coal is a fossil fuel which is mostly comprised of carbon and small amounts of hydrogen, sulfur, oxygen, and nitrogen, heating it in the absence of air (the so-called "destructive distillation of coal") yields plenty of useful products including coal gas, coal tar, and coke. These by-products were first thought to be useless. Indeed, the coal tar, the large amount of oily tar produced by the distillation, was considered so worthless that anyone could have it for free, even though millions of tons of coal were expended every year to by-produce the stuff.

Slowly, though, a different hydrocarbon reality emerged. Chemists became able to extract useful chemicals from the coal tar. Then, a revolution occurred in 1856. A chemist born in the East End of London serendipitously discovered that he could isolate a beautiful purple molecule from the coal tar. William Henry

Perkin, in laying the foundation for the synthetic organic chemicals industry, helped revolutionize the world of fashion.

The revolution that would reshape the global economy truly had Perkin as its catalyst. In 1856, during Easter vacation at the tender age of eighteen, Perkin was perched in his attic laboratory in his family's East End home. He was using his vacation to dabble in experiments with coal tar at the suggestion of his mentor at the Royal College of Chemistry, August Wilhelm von Hofmann. Perkin took up Hofmann's challenge and, in a series of experiments, stunned the world.

In one revelatory iteration, Perkin washed the black goo out of a test tube to reveal something that startled him: a vivid purple residue on the glass. Not only was this substance's color stubborn, but its bright tincture also transferred flawlessly to a cotton cloth he used to clean his test tubes. Perkin didn't know *why* the resulting color was so vivid; a chemical's capacity to absorb light at certain wavelengths based on the structure of their shared electron bonds wouldn't be understood for another half century. Nor did he know the chemical he had created; the exact molecular formulation of his new substance wouldn't be fully worked out until the 1990s.

But Perkin wasn't dumb. Just a few months before, he and a fellow student tried to synthesize a textile dye and failed; now, he had somehow succeeded while doing Hofmann's bidding to try creating a medicine for malaria. Perkin could see that this new and vivid substance at the bottom of his test tube might prove not only useful, but also very lucrative. Whoever made the world's first dye, capable of staining cotton, silk, and many other textiles with a beautiful color, might get very rich indeed.

Perkin would also have been aware of the ancient thalassocratic civilization known as the Phoenicians. Originating in the Levant region of the eastern Mediterranean, and from humble beginnings, Phoenician civilization eventually extended throughout the Mediterranean from Cyprus to the Iberian

Peninsula. The Phoenicians came to prominence in the mid-twelfth century BC, following the decline of the most influential cultures in the Late Bronze Age collapse. Renowned among contemporaries as skilled traders and mariners, becoming the dominant commercial power for much of classical antiquity, their name *Phoenicia* most likely described one of their most famous exports, a dye also known as Tyrian purple. This was a secretion extracted from several species of predatory sea snails in the family Muricidae, especially the spiny dye-murex, found in the Eastern Mediterranean Sea. As a result, the color purple had long been associated with royalty and nobility. Tyrian purple was extremely expensive in antiquity (each murex produced only a few drops of dye, and only when freshly caught, so it took eight thousand snails to produce one gram of Tyrian Purple; such extravagance had given rise to the word "porphyriogenatos," which literally means "born in the purple"). Purple became the color of choice worn by Roman magistrates. Julius Caesar first decreed that only the emperor and his family could wear purple garments. It did him little good. On the Ides of March in 44 BC, Caesar was wearing his ceremonial robe of Tyrian purple when he was assassinated by Brutus in the Roman senate. Thirteen years later, at the Battle of Actium, the sails of Cleopatra's royal barge were dyed vivid purple. It was the imperial color worn by the rulers of the Byzantine Empire, the Holy Roman Empire, and later by Roman Catholic bishops.

But when the Roman Empire declined and died, the system of murex cultivation and dye production died along with it. The purple of Caesar and Cleopatra, hue of wealth and power, king of all colors, no longer lingered in the dye maker's palette. It became legend. Yet, here it was, clinging tightly to the glass walls of Perkin's test tubes. Perhaps, the teenager thought, his failed experiment might not be such a failure after all. In time, everything fell into place. With hard work and a fair lick of luck, Perkin became rich.

Given this scientific context of Perkin's astounding discovery, it should come as no surprise that during the period after *The Final Problem*, Conan Doyle had Sherlock work on coal tar derivatives in Montpellier, a city in southern France near the Mediterranean Sea. We never learn what aspect of chemical research into coal tar derivatives Sherlock was carrying out. Some suggest he was trying to separate carcinogens from coal tar, and others that Sherlock was working on the development of radiation technology. But, given the Perkin context, the consensus is that Sherlock's research was on synthetic dyes.

During the days of Sherlock's "Great Hiatus" (1893 to 1903), Britain was losing ground in capital's international chemical competition to see who would dominate the dyes market. Perkin made a splendid start for the British. He secured financial backing from his father, built a dye factory, and had swiftly become wealthy, but Perkin retired from the industry at age thirty-six. It was the first time folks realized that the study of chemistry could make them rich.

Chemists in other countries began to build their own dye factories. The industry flourished. The British, Perkin apart, seemed slow to catch on, as their scientific establishment had an old-fashioned attitude to the commercial aspects of the craft. Not so for the Germans, who smartly pursued the profits to be drawn from dyes. It wasn't long before the German dye industry bettered that of Britain. Soon, 80 percent of dyes bought in Britain were being manufactured in Germany. Little wonder that the celebrated British chemist Henry Roscoe would complain in 1881, "To Englishmen it is a somewhat mortifying reflection that whilst the raw materials from which all these coal tar colors are made are produced in our country, the finished and valuable colors are nearly all manufactured in Germany."

Watson described London in the first pages of *A Study in Scarlet* as "that great cesspool into which all the loungers and idlers of the Empire are irresistibly drained," so the consensus idea that Sherlock's chemical research was on synthetic dyes makes total sense. In effect, Doyle has Sherlock engaged in a patriotic attempt to revive the British dye industry. Along with *The Sign of Four*, Conan Doyle's first two novels also aim to establish concerns about Britain's place in a borderless world. Half of *A Study in Scarlet* is about matters in the United States, and a fair chunk of *The Sign of Four* is about a plot to steal Indian treasure which leaves a long trail of murder. Rather than being constrained to the borders of America and India, these crimes transcend national boundaries and become woven into Sherlock's case investigations on British soil.

Indeed, a strikingly outlandish array of characters in Doyle's detective tales return to Britain after a period in the colonies. There's the buckled and bilious ex-trooper, owner of a pet Indian mongoose. There's a one-legged chap (crocodile got the other) with the poison-toting pygmy companion. There's a fellow who keeps a ferocious hound and claims his South American wife is his sister. And there's a dubious character who returns from South Africa with a "blanched" face and a furtive aspect. Doyle portrays these figures as brooding and sinister, and their return to Britain triggers crime and crisis. Meanwhile, in actual fact, folks returning from the British colonies to London was a regular affair, and such ex-colonials were familiar figures in Victorian society. So, we might ask, why does Doyle depict such returnees as problematic, apart from the obvious expedient of creating suspense?

English historian Eric Hobsbawm has argued that it wasn't until the last decade of the nineteenth century that the nation became a political entity and nationalism an ideological force. In this context, we can interpret Sherlock as a popular agent who plays a valuable role in policing the borders of Britain's emerging nation. Sherlock is not just the quintessential detective, but also the quintessential Englishman. (Doyle uses Sherlock to underline a register of core bourgeois English values: nationalism, property, the

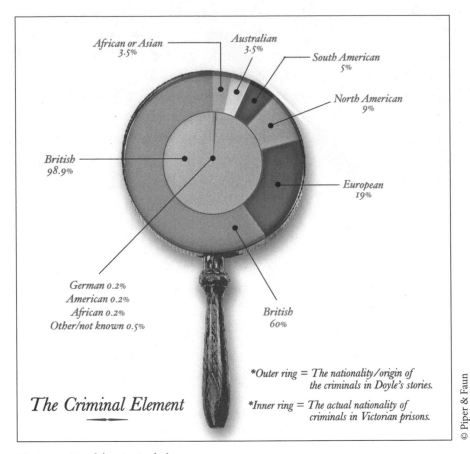

African or Asian
3.5%

Australian
3.5%

South American
5%

North American
9%

British
98.9%

European
19%

German 0.2%
American 0.2%
African 0.2%
Other/not known 0.5%

British
60%

*Outer ring = The nationality/origin of
the criminals in Doyle's stories.

*Inner ring = The actual nationality of
criminals in Victorian prisons.

The Criminal Element

© Piper & Faun

Figure 6. Doyle's criminal element.

laws of exchange, and the sanctity of wedding vows. Yet Sherlock
is an odd choice to defend such values. He is an asexual drug user
who exhibits deeply impolite and antisocial behavior!)

Sherlock's silhouette is now recognized all over the world. But,
most importantly for our argument here, Sherlock has been and
is still used to market and boost business for Britain. Of course,
Doyle would use Sherlock as a signifier to try reviving the British
dye industry. Conan Doyle himself has been dubbed, in Jon
Thompson's *Fiction, Crime, and Empire,* "one of the great Victorian
apologists of empire." After all, Doyle was knighted for his writings
in support of Britain's actions in *The Great Boer War,* and he wrote

in his *Memories and Adventures* about his adamant belief in the nation-building talents of Anglo-Saxons abroad. But charity begins at home, and Doyle must also look after Britain's interests.

Naturally, the German view on the state of the international dye trade was the total opposite to that of British chemist Henry Roscoe. For example, the German chemist and hygienist Theodor Weyl wrote in the preface to his 1885 book on coal tar colors, "Thanks to the cooperation of theory and practice, the coal tar industry of Germany has conquered the world, and inasmuch as new and improved methods are continually being devised, will be able to maintain its pre-eminent position." Doyle no doubt knew about the decline in British dye, so he had Sherlock work on coal tar derivatives with German dominance in mind. In short, Sherlock, the chemistry virtuoso, was doing research in an attempt to turn the tide of German industry.

Poisons That Open Your Eyes

In Guy Ritchie's 2011 film *Sherlock Holmes: A Game of Shadows*, Sherlock easily identifies the poison on a dart that had instantly paralyzed its victim. How? By simply sniffing it. "Curare!" he declares. Curare really is a poison, but identifying it by smell alone belongs only in film and fiction. Sir Walter Raleigh, English statesman, soldier, writer, explorer, and a favorite courtier of Queen Elizabeth I, is credited with first bringing a sample of curare to Europe, where it was christened "curare" based on the word *wurari* from the Carib language of the Macusi of Guyana.

Long before Europeans arrived, the wise indigenous peoples in Central and South America used curare to tip their arrows as a paralyzing agent for hunting, and for therapeutic purposes. The substance only becomes active by a direct wound contamination by poison dart or arrow. Lamentably, little scientific attention was paid to curare until 1812 when English naturalist Charles Waterton discovered that, if the dose were right, nonfatal muscle relaxation could be accomplished. So it was that curare came to be used in the treatment of infantile paralysis, lockjaw, and even epilepsy.

The scene in Ritchie's *Game of Shadows* isn't based directly on any of Doyle's classic tales. Various references from the Sherlock stories are merely weaved into the movie. But curare does play a role in *The Adventure of the Sussex Vampire*, just one example of a Dr. Doyle tale in which poisons are incorporated. As progress in chemistry advanced during the nineteenth century, so did the methods of working with different substances, and the means of distinguishing between them became more refined. This was true, too, of toxins and poisons.

As a medical doctor, Doyle would have been well aware of the use and misuse of drugs. Take chloroform, for instance. Now known as trichloromethane, chloroform is a strong-smelling, colorless, and dense liquid which, when inhaled or ingested, is a very powerful anesthetic, sedative, or euphoriant. How do we know? Because Sir James Young Simpson, whose alma mater, like Doyle's, was the University of Edinburgh, deliberately inhaled chloroform to see if it had anesthetic properties. Simpson, a Scottish obstetrician and father figure in the history of medicine, became the first physician to demonstrate the anesthetic properties of chloroform on humans and helped to popularize its use in medicine. His ingenious plan was simply to drink the stuff. On the evening of November 4, 1847, along with two assistants (it's not clear whether they had any choice in the matter), Simpson inhaled a selection of chemicals to check their properties. They gave benzene a bash, which is pretty daring, as benzene is a toxic and flammable liquid now classified as a carcinogen that increases the risk of cancer and is also a notorious cause of bone marrow failure. Next up was acetone. Again, a flammable liquid, with a characteristically pungent smell and now known as the active ingredient in nail polish remover and paint thinner.

Could chloroform be used as a poison? Certainly. The substance has something of an early history as a murder weapon, and sometimes chloroform played a role in real tragedy. In 1911, a Long Island father killed his son and two daughters with chloroform and then, leaving a suicide note, walked away into the gray Atlantic. A less tragic but more gruesome use of chloroform happened in Yonkers, New York, four years later. Frederic Mors, a recent immigrant from Vienna, used the substance to murder elderly pensioners at the German Odd Fellows home in Yonkers. Mors was allowed to practice his medical interests when he was asked to nurse the men, as Mors was better educated than most orderlies, but the home's superintendent soon asked Mors to take on another job, to help with the "removal" of some of the sickliest (and costliest) residents by administering whatever dose of chloroform did the trick.

Mors first tried arsenic but the elderly victim chosen for this first substance experiment didn't die as desired. He was inconveniently and messily sick, developing a kind of creeping paralysis, living on for several wretched days, so Mors returned to the dispensary. Seduced by the sweet chemical sting of chloroform, he tried his next choice. The effect was so pleasing and convenient that Mors later confessed, "When you give an old person chloroform, it's like putting a child to sleep." Mors liked chloroform for its efficiency. He went on to murder seven more residents of the home with little trouble. Until he was caught. It's a wonderful poison, really, he confessed, perhaps a little cloying in its oversweet smell, but superbly, reliably lethal.

When Frederic Mors eased that vial of chloroform off the dispensary shelf, it was still mostly used and known for being miraculous rather than murderous. Conan Doyle uses chloroform in three Sherlock tales, but its use never results in death. One of them, *The Disappearance of Lady Frances Carfax*, published late in 1911, came after the Long Island case. The other two, *His Last Bow* and *The Adventure of the Three Gables*, came after the Frederic Mors murders.

It's the *Lady Frances Carfax* tale in which chloroform is most dramatically used. Over a month has gone by since the wealthy Lady Frances last wrote home from Lausanne in Switzerland, so Watson goes off to investigate, and finds her in the Englischer Hof hotel in Baden-Baden. It seems that a vicious rascal of an Australian by the name of Henry Peters (his earlobe was chewed off in a bar brawl, for heaven's sake; how badass is that?) steals the Lady's jewelry, carts her off to his home in London, then attempts to murder her by burying her alive in an occupied coffin. This is where the chloroform comes in. She's sedated in the coffin. Sherlock manages to track down Lady Frances when her jewelry is pawned, gets to the scene in time to fathom the occupied coffin stunt, but not in time to catch the badass Australian.

Meanwhile, in *The Adventure of the Three Gables*, Barney Stockdale is hired by a wealthy woman, Isadora Klein, who is used to getting what she wants. Stockdale uses chloroform to subdue Mary Maberley, a lady who lives at Three Gables, a house at Harrow Weald. The burglars then steal a manuscript written by her son Douglas. Why? Klein is determined to get the manuscript, as its publication will reveal her past and undoubtedly result in cancellation of her impending marriage to the young Duke of Lomond. When Sherlock arrives at the scene, Klein has burned the manuscript, which prevents Sherlock from returning the manuscript to Mrs. Maberley. Not to be denied, and as compensation, Sherlock cajoles Klein into bankrolling a first-class trip around the world for Mrs. Maberley.

Finally, in *His Last Bow*, Sherlock plays the part of a double agent to great effect. He seems to be working to get British naval secrets for the German spy Von Bork, convincing Von Bork that he is an American by the name of Altamont. Sherlock is driven by his chauffeur to deliver the material, but the chauffeur is Watson. Together, the two of them subdue the German, who is chloroformed and captured. Set in 1914, *His Last Bow* is poised at the outset of World War I.

Dr. Doyle mentions other poisons throughout the sixty Sherlock stories, but he rarely opts for the more sensational options. For example, in spite of its infamy as the "inheritance powder," Doyle never uses arsenic in any Sherlock case, though Aqua Tofana, a colorless and tasteless elixir of mostly arsenic, lead, and maybe belladonna, reportedly created in Sicily around 1630, gets a passing reference in Doyle's first tale. Aqua Tofana is associated with Giulia Tofana, a professional poisoner from Palermo who was said to be the ringleader of six poisoners in Rome that sold Aqua Tofana to would-be widows and used it to commit over six hundred murders. The legend that Mozart was poisoned using Aqua Tofana is apocryphal, even though it was Mozart himself who began the rumor.

The Hound of the Baskervilles

As stories go, *The Hound of the Baskervilles* checks a lot of boxes. The third of Doyle's four novels, *Hound* was Sherlock's first appearance since his "death" in *The Final Problem* almost ten years before. And what a return to form it was. Set in the darkest Devon of Dartmoor, our dynamic duo is on the trail of an attempted murder inspired by the legend of a terrifying, diabolical hound of supernatural origin. One of the most famous tales ever told, in 2003 the book was ranked at number 128 of 200 on the BBC's *The Big Read* poll of Britain's best-loved novels. In 1999, a poll of Sherlockians ranked *Hound* as the best of the four Sherlock novels.

In terms of our Deduction Diamond, one might wonder where on earth (or off it!) Conan Doyle got the idea for such a fantastic tale. In short, what's this text's context? Writers are often asked from where their ideas spring. Some plots come from newspapers, perhaps, or true crime tales. Others originate simply from their fertile imaginations. But maybe most often, stories are sparked from an anecdote told by a friend, and not necessarily with the intent of its being used as the basis of a novel. Unless the friend being told the tale is Sir Arthur Conan Doyle.

So it was with Conan Doyle and his friend Bertram Fletcher Robinson, journalist and folklorist. Robinson entertained Doyle with horror stories of Devon while they were together on a golfing holiday (it seems that Doyle, like Sherlock, was a fan of sensational stories). The tales told included legends of the spectral hounds that were rumored to roam Dartmoor, phantom beasts with red eyes, whose huntsman is Lucifer, Lord of Light. Doyle was forever grateful for the idea. "My dear Robinson," Doyle wrote later, "It was your account of a west country legend which first suggested the idea of this little tale to my mind. For this, and for the help which you gave me in its evolution, all thanks. Yours most truly, A. Conan Doyle."

Consider Lucifer, "Lord of Light." In Roman folklore, Lucifer meant "light bringer" in the native Latin. But Lucifer was also the name given to the planet Venus, though it was often personified as a male god bearing a torch. The Greek name for the same planet was sometimes Heosphoros, which means "dawn bringer," or Phosphoros, meaning "light bringer." Phosphorus, as it's spelled in English, is a beautiful word. And the element which bears that name is everywhere in nature. It's found in the brain, and it's found in fish. (So, eating fish makes you smarter!) It's also found, of course, in the tips of matches and in will-o'-the-wisps, that atmospheric ghostly light seen by travelers at night.

Framed in Flickering Flame

It was a ghostly light that framed the diabolical hound in Doyle's *Hound of the Baskervilles*. Spied at night on Dartmoor's dark and brooding landscape, Watson says of the hound, "Its muzzle and hackles and dewlap were outlined in flickering flame." The hound hunting down Sir Henry Baskerville was "not a pure bloodhound and not a pure mastiff; but it appeared to be a combination of the two; gaunt, savage, and as large as a . . . lioness." According to legend, a curse ran in the Baskervilles since the time of the bloody English Civil War. Sir Hugo Baskerville's abduction of a maiden on the moor had led to the poor girl's death. But there was a price to pay. The "wild, profane and godless" Sir Hugo was himself murdered on the moor by a huge demonic hound. Allegedly, the same creature had been haunting the manor ever since, causing the premature death of many Baskerville heirs. Of those who observed it, "One, it is said, died that very night of what he had seen, and the other twain were but broken men for the rest of their days." And so a curse was laid upon all future heads of the house of Baskerville (the name Baskerville was perhaps taken from Harry Baskerville, the coachman who had driven Conan Doyle and Robinson's pony and trap about Dartmoor).

Doyle's Technique

It's not just the story setup that makes *Hound of the Baskervilles* one of the most famous tales ever told. In our Deduction Diamond, recall that one of the four points of the Diamond is named techniques and refers to the craft Conan Doyle used to create the effects on his readers. So, what techniques did Doyle employ to make this tale so esteemed, so memorable?

Well, it's known that Doyle was a writer who relied to some extent on formula. The brilliant-but-broken formula of Sherlock's character. The monthly episode formula that Doyle's reading public so adored. Another featured formula in Doyle's fiction was to bring his protagonist back to his homeland after half a lifetime in foreign climes. These countries would include exotic locations like America, India, Australia, or South Africa. This gave Doyle the chance to make his main character a gold prospector, say, or an adventurer of some sort. A typical protagonist would return home to mete out justice on the person who had been the root cause of their exile, all those years before.

Enter the young Sir Henry Baskerville. He appears to have lived most of his young life (he's around thirty years of age) in America and Canada. But, due to the deaths of a couple of potential heirs, Sir Henry is now the owner of Baskerville Hall. His is "the serious and extraordinary problem" Sherlock loves to discover in one of his detective cases. Sir Henry was warned of the dark legend of the hound, and his likely fate, should he be determined to "go to the home of his fathers," a huge and brooding pile, set in the middle of Dartmoor.

Doyle sets the scene of how vast and gloomy the seat of the Baskervilles is. He writes:

> The avenue opened into a broad expanse of turf, and the house lay before us. In the fading light I could see that the center was a heavy block of a building, from which a porch projected. The whole front was draped in ivy, with a patch clipped here and there where a window or a coat of arms broke through the dark veil. From this central block rose the twin towers, ancient, crenellated and pierced with many loopholes . . . A dull light shone through heavy mullioned windows, and from the high chimneys which rose from the steep, high-angled roof there sprang a single black column of smoke.

You can see how expertly Doyle sets a scene, thick with menace and foreboding, in which the drama will soon unfold. The story is the same indoors:

> The dining-room which opened out of the hall was a place of shadow and gloom. It was a long chamber with a step separating the dais where the family sat from the lower portion reserved for their dependents. At one end a minstrel's gallery overlooked it. Black beams shot across our heads,

with a smoke-darkened ceiling beyond them. With rows of flaring torches to light it up, and the color and rude hilarity of an old-time banquet, it might have softened: but now, when two black-clothed gentlemen sat in the little circle of light thrown by a shaded lamp, one's voice became hushed and one's spirit subdued.

Doyle doesn't overdo it. He doesn't let his description of the house and Dartmoor get too bogged down with doom, wicked curses, and blood-chilling dread, as some lesser writers might have done. No, Doyle doesn't gild his lily. Rather, his narrative is nuanced with descriptions of the countryside with loving care and finesse, perhaps more than any other of his tales. He paints a picture of the "red earth new turned by the plough and the broad tangle of the woodlands." He describes how "all was sweet and mellow and peaceful in the golden evening light" and talks of the "green squares of the fields and the low curve of a wood" from which there "rose in the distance a grey, melancholy hill, with a strange jagged summit, dim and vague . . . like some fantastic landscape in a dream." Later, when he describes "black tors," "craggy summits," and "melancholy downs," the contrast is all the more dramatic.

The Beast of the Baskervilles

What of the hound itself? How does Conan Doyle build the reader's expectation of what's to come? Doyle describes it as "a foul thing . . . a great black beast, shaped like a hound, yet larger than any hound that ever mortal eye has rested upon." Not only that, but it was "luminous and ghastly" with "dripping jaws" and "blazing eyes." Thinking about the effects point of the Deduction Diamond, we might wonder what Doyle's text makes his readers feel or think about his descriptions. The answer is pretty clear from Sidney Paget's illustrations in *The Strand Magazine*. Though the creature is shown leaping out of its hiding place to savage its victim, it hardly fits the Irish wolfhound description in Doyle's text. The dog in Paget's drawing is surely a German Shepherd. In those genteel days of the early twentieth century, before movie monsters such as Nosferatu, Pennywise, or Pumpkinhead, folks were easily scared, so we can assume that Doyle's words had the desired effects without further amplification from Paget.

Needless to say, as a fine upstanding and fearless member of the British Empire, the intrepid young Sir Henry is not put off by the curse of the creature destined to plague Baskerville heirs. Nor is he daunted by the mysterious death of the previous occupant of Baskerville Hall. Like a good rationalist, Sir Henry seeks a credible explanation for Sir Charles's death. Charles had a dickie heart. He died from a timely exhaustion of the old ticker; given the proximity of a vicious "Baskerville demon," the old boy could hardly be blamed.

Sherlock isn't taken in at all, of course. Sure, there was some extraordinary witness evidence. All the superstitious locals agreed that the hound was huge, and it was luminous, ghastly, and spectral, too. Yet, from the get-go, Doyle's readers knew that this creature "could not possibly be any animal known to science." Even if it had brought about fear and trepidation in the district. Even if the bravest had baulked at the idea of crossing the moor at night.

No, Sherlock knew some human agency was at work. At first, Sherlock alludes to devilry. This inspires Watson to ask if Sherlock favors some kind of supernatural cause, but Sherlock will have none of it. "The devil's agents," he says, "may be of flesh and blood, may they not?" Indeed, nowhere in Sherlock's tales is a single spirit, ghost, or demon to be discovered. Doyle simply doesn't allow them to intervene. That's just what one might expect from a writer of relatively modern fiction, and it's certainly what one might expect from the science virtuoso Sherlock Holmes, but Doyle himself actually later believed in paranormal phenomena. He went to spiritualist meetings and séances. He took photographs of beings he believed to be fairies. (Something within Doyle had changed at the end of WWI. He seemed like a different man, marked by the horrors of war and the death of his eldest son. He began venturing into fringe territories. Once an avowed atheist and militant rationalist, Doyle became a passionate supporter of spiritualism.) Yet, he never allowed such fancies to influence his detective fiction. We can only assume that Doyle was so deeply devoted to the precise portrayal of his most famous character that, when writing of Sherlock, Doyle was only able to let Sherlock believe in what Sherlock would believe, and not express any convictions alien to his own.

That Phosphoric Hound

What of the hound's "fearsome glow?" Watson decides that the element phosphorus may have been put around the mouth of the dog because phosphorus, when exposed to air, glows in the dark, but Sherlock is skeptical. As a practicing chemist, Sherlock knows that there is no odor from whatever chemical was used, so that nothing would interfere with the hound's sense of smell, so Sherlock concludes that "a cunning preparation" must have been used.

Who is scientifically correct? Watson, the physician, who holds that phosphorus is the cause? Or Sherlock, the skeptical chemist? As we have seen throughout this chapter, when it comes to a chemical phenomenon in Doyle's texts, Sherlock is almost always right. Anyone who considers chemical analysis relaxing (I have a degree in chemistry, so please believe me, it's *not*!) is clearly a chemistry devotee. Sherlock adored his chemistry, and often got so absorbed in his analysis that he worked late into the night.

The fearsome glow of the dog was more likely due to chemiluminescence, not phosphorescence. A chemical reaction using something other than phosphorus outlined that hound in flickering flame. After all, what dog could bear to have phosphorus slobbered around its muzzle? Mind you, this effect might certainly have driven the dog to murder! Whatever chemical was used on the hound, it certainly did the trick. Not only did it result in a terrifying apparition, but it also took Sherlock five shots to kill stone dead one of literature's most enduring dogs.

Hounds of Baskerville

Enduring, that is, at least until a century later when the BBC's crime drama television series *Sherlock* adapted Doyle's tale in their episode, "The Hounds of Baskerville." There's still a guy called Henry, though not a Baskerville, and he witnessed his father's death by a "gigantic hound" on Dartmoor twenty years before, prompting his request for Sherlock to take up his case.

At the hands of the creative team at the BBC, Baskerville has now become a Ministry of Defense research facility (some of the exterior scenes at Baskerville were filmed at a gas works a mere six miles away from my home). "H.O.U.N.D." is a covert CIA project whose aim is to synthesize an antipersonnel chemical weapon that is capable of inducing hallucinatory effects on its victims. When Sherlock cracks the case, it turns out that Henry's father was murdered by a rogue scientist because he found him testing the dubious drug.

"The Hounds of Baskerville" still manages to cram in an actual dog. The local innkeepers kept a dog on the moor to boost the tourist trade and those who saw it, influenced as they were by H.O.U.N.D. hallucinogen, believed it to be monstrous. Ingenious.

CHAPTER SIX

AN ADVENTURE IN MISCELLANY AND MURDER

. . . in which we examine Sherlock's use and knowledge of the other sciences, from algebra to further physics.

Watson: I picked up a magazine . . . and I naturally began to run my eye through . . . "The Book of Life" . . . it attempted to show how much an observant man might learn by an accurate and systematic examination of all that came in his way. It struck me as being a remarkable mixture of shrewdness and of absurdity. The reasoning was close and intense, but the deductions appeared to me to be far-fetched and exaggerated. The writer claimed by a momentary expression, a twitch of a muscle or a glance of an eye, to fathom a man's inmost thoughts. Deceit, according to him, was an impossibility in the case of one trained to observation and analysis. His conclusions were as infallible as so many propositions of Euclid. So startling would his results appear to the uninitiated that until they learned the processes by which he had arrived at them they might well consider him as a necromancer.

—Sir Arthur Conan Doyle, *A Study in Scarlet* (1887)

Sherlock and Lenses

We know that Sherlock was king of chemistry, but he was very knowledgeable in the other sciences too. Much earlier in this book we quoted an excerpt from *A Study in Scarlet* in which Watson lists what he perceives to be the knowledgebase of his newfound friend, Sherlock. You'll hopefully recall that, whereas Sherlock had a profound knowledge in some subjects, obviously including chemistry, he was thought to be quite ignorant in others. The one glaring lack in that knowledgebase, according to Watson, was the Copernican Theory and the composition of the Solar System. In short, as Watson put it, as far as astronomy went, the sum total of Sherlock's knowledge was "nil."

Optics is that branch of physics in which the behavior and properties of light are analyzed. Optical astronomy is the study of heavenly objects using telescopes in visible light, about which Sherlock knew next to nothing. Yet Sherlock's use and knowledge of another kind of optics, namely the magnifying glass, is quite profound. It is one of Sherlock's prime tools of deduction.

Readers and viewers alike commonly associate Sherlock with his trusty magnifying glass. The telescope was often known in the early days as the "far-seer." In which case, Sherlock's magnifying glass qualifies as a "near-seer." Optical instruments have fascinating associations from the early days of the search for extraterrestrial life, ironically enough for Sherlock's lack of awareness of astronomy. Medieval Dutch eyeglass makers were at the forefront in the development of early optical instruments. The microscope, a more technical "near-seer," unveiled evidence of previously unknown, though minuscule, life. The implication was clear. Beyond normal human perception lay remote worlds. These small and remote worlds were put to astronomy's grand cause in the search for life in space. Might not the cosmos at large also harbor unseen worlds? Not that Sherlock was bothered!

In fact, a telescope is indeed used in a Sherlock tale, but not by Sherlock himself, and not to look at the stars. The story concerned is *The Hound of the Baskervilles*, and the telescope-wielder is one of Sir Henry's gentleman neighbors, Mr. Frankland. What sets Frankland apart from the other townspeople is his almost constant obsession with snooping on his neighbors. And what better instrument to snoop with than a telescope. Frankland uses the "far-seer" to keep a beady eye on all activities on the moor. When Watson looks through the 'scope, he confirms Frankland's suspicion of dubious activity, and at once goes out on the moor to investigate. Watson is somewhat stunned to find that the suspicious character living out on the moor is none other than Sherlock himself. Sherlock and Watson had been working independently on the Baskerville case but, now that Sherlock's game is up, they work together.

The magnifying glass is one of several optical devices used in the Sherlock tales and is referenced in twenty of the sixty stories. From the get-go, in *A Study in Scarlet*, Sherlock devotes a full twenty minutes to scouring the room where the cadaver of Enoch Drebber is found. Lens in hand, Sherlock is described by Watson as "sometimes stopping, occasionally kneeling, and once lying flat upon his face." It's the same in the next tale, *The Sign of Four*. Here, Sherlock makes expert use of his magnifying glass in examining the watch belonging to Watson's brother, pouring over the timepiece with a convex lens, and coming to those deductions that so startled his companion. Later in the same story, Sherlock uses the glass to look at the rope which Jonathon Small climbed to get into Bartholomew Sholto's room. Again, in *The Sign of Four*, "he whipped out his lens and a tape measure and hurried about the room on his knees, measuring, comparing, examining with his long thin nose only a few inches from the planks." In short, Sherlock uses his magnifying glass to study the murder room.

Sherlock's magnifying glass is as iconic as his deerstalker hat, his Calabash pipe, and his violin, irrespective of the provenance of those familiar items. Passages such as this from *The Boscombe Valley Mystery* are key to how we think of Sherlock the science virtuoso in action:

His face flushed and darkened. His brows were drawn into two hard black lines, while his eyes shone from beneath them with a steely glitter. His face was bent downward, his shoulders bowed, his lips compressed, and the veins stood out like a whipcord in his long sinewy neck. His nostrils seemed to dilate with a purely animal lust for the chase.

Sherlock uses a microscope in *The Adventure of Shoscombe Old Place*, the last Sherlock tale published by Conan Doyle. Though his microscope work pays dividends, *Shoscombe Old Place* is one of those tangential tales that we never get to hear about. But, as *Shoscombe Old Place* is Doyle's last tale, we can welcome Sherlock's use of the microscope as evidence that he was progressing as a forensic detective.

The BBC production *Sherlock* took care to show this progression. In a number of episodes, we actually see Sherlock *using* a microscope. For example, we previously mentioned the episode "The Reichenbach Fall." This is the story in which, among other things, Sherlock and John investigate the kidnapping of the children of the British Ambassador to the US. When Sherlock uses his skills in chemical analysis to identify an array of contact materials caused by foot printing, we see him busily at work at the microscope.

Sherlock and the Sky

Doyle's tales also provide evidence of an evolution in Sherlock's understanding of astronomy. Sherlock did indeed come to care about the Solar System. This is despite his sum total of knowledge being "nil" when we are first introduced to Sherlock by Watson in *A Study in Scarlet*. It would seem very unlikely that a science virtuoso such as Sherlock would be unaware of Copernican Theory since the Copernican Revolution was a revolution of ideas, a transformation in our very conception of the Universe and our relation to it, but let's forgive Doyle and assume that early Sherlock was simply a far more practical character.

By the time Doyle is writing *The Bruce-Partington Plans*, Sherlock is well acquainted with the Solar System. The scene is this: Sherlock is stunned to learn from an urgent telegram from Mycroft that his brother is about to imminently blow in at Baker Street. With the Solar System no longer a mystery, Sherlock remarks that for the famously lackadaisical Mycroft to leave the luxury of his Diogenes Club and roll up at Baker Street lodgings is somewhat like a planet being ripped out of the seat of its orbit. Given that Sherlock is well-versed in astronomy by this point, the forty-second story, we need to look deeper into Doyle's tales to find clues of Sherlock's astronomical development.

We find a first clue in *The Adventure of the Musgrave Ritual*, the twentieth story. In this tale, Sherlock calculates the correct position of the Sun for his geometrical reckoning of where the shadow of an oak tree will fall. He feels that he will not need to allow for "the personal equation as the astronomers have dubbed it," a clear reference to the fact that Sherlock had been doing some reading in the field of astronomy.

By the time we get to the twenty-fourth story, *The Adventure of the Greek Interpreter*, Sherlock has gotten pretty technical. Doyle treats us to a discussion between Sherlock and Watson on the "obliquity of the ecliptic." Now, the *ecliptic* is the plane of Earth's orbit around the Sun. That needs some unpacking. It's not easy, building a true mental model of the sky. That, in itself, shows how much Sherlock has developed his astronomy.

Consider this scientific context of our Deduction Diamond. In luxurious comfort, Sherlock sits at the center of a transparent Earth. A gargantuan glass globe, clear as crystal. Comfort is crucial, as Sherlock needs this to make his usual painstaking observations. Way above Sherlock's head is a collection of airborne gods of the ancient Greek world (this is, after all, *The Adventure of the Greek Interpreter*). Each god is flying along a concentric orbit, and each represents one of the wandering planets. Sherlock sits and watches the divine traffic. Each god appears as a luminous point, and each moves at a different speed along a narrow lane known as the "zodiacal belt." The great glass globe begins to move. It rotates around Sherlock as he remains at rest. The direction of the spin of the globe and the flow of the traffic is one and the same. The sphere and the gods rotate

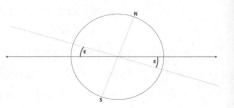

in the same direction, but the holy traffic remains confined to the "zodiac." In this glass globe scenario, the ecliptic is an imaginary line in the sky that is the plane of Earth's orbit projected out in all directions. This reference plane, the ecliptic, also marks the apparent path of the Sun, as the Sun is the focus of Earth's orbit. The Moon and wandering planets also travel along the path of the ecliptic, so the ecliptic is the projection of Earth's orbit into local space. It's also used to mark the very plane of the Solar System. Tracing the paths of the planets in front of the background stars, the ecliptic passes through the constellations of the zodiac.

But what about the "obliquity of the ecliptic" that Sherlock mentions in *The Adventure of the Greek Interpreter*? This is the angle between the Earth's equator and the ecliptic. It has the same value as the Earth's axial tilt. If the Earth were not tilted, but had its axis of rotation perfectly upright, the obliquity of the ecliptic would be zero. In reality, the average obliquity is just over 23.6°, but this value decreases slowly per century because of planetary perturbations of the Earth's orbit. In fact, the obliquity oscillates between limits of about 22° and 24.5° with a mean period of some 41,000 years. In the current cycle it was at a maximum of 24.2° around 9,500 years ago and will reach a minimum of 22.6° in another 10,200 years before starting to increase again. See what we mean about Sherlock's astronomy knowledge now being quite technical? He clearly knew enough astronomy that he could participate in such a conversation.

One last astronomy topic that pops up in Doyle's Sherlock tales concerns that diabolical astronomer, Professor Moriarty. We learn in *The Final Problem* that Moriarty had secured a chair in mathematics for his paper, "Treatise on the Binomial Theorem," yet his most impressive work was astronomical. *The Dynamics of an Asteroid* was apparently a book "which ascends to such rarefied heights of pure mathematics" that few could even read it. (We briefly see the cover of *The Dynamics of an Asteroid* in the 2011 movie, *Sherlock Holmes: A Game of Shadows*.)

It seems that Moriarty had developed an interest in astronomy once he got tenure. Indeed, this inclination even appears to survive his becoming a London crime lord. In the fourth and final Sherlock novel, *The Valley of Fear*, published in 1915, Inspector MacDonald calls upon Moriarty to question him. Moriarty can't help but tell the inspector all he needs to know about eclipses, even providing a demo on how they happen, and lending MacDonald a book on the subject.

Astronomy's fascination with asteroids had begun a century before. During the 1700s, astronomers were intrigued by a mathematical equation known as Bode's Law. (Johann Elert Bode was the German astronomer who also helped determine the orbit of Uranus and suggested the planet's name.) The Law seemed to predict the orbital locations of the known planets, but with one exception: There should be a planet between Mars and Jupiter. In 1781, Sir William Herschel discovered Uranus, the seventh planet, and it was at a distance that corresponded to Bode's estimate. Excitement about the Law then reached a peak, with many astronomers totally convinced that a planet must exist between Mars and Jupiter.

The end of the 1700s saw the emergence of the "Celestial Police." This was the name adopted by a group of astronomers banded together at the observatory at Lilienthal, Germany, to hunt down the missing planet. But they were beaten to the punch on New Year's Day, 1801, when Italian Catholic priest Giuseppe Piazzi of the Palermo Observatory discovered what he believed to be the missing planet.

Piazzi discovered the first asteroid, Ceres. Subsequent observations established that Ceres couldn't be the missing planet. With a diameter of just 587 miles, it was simply too small. Instead, it was classified as a "minor planet" and the search for the missing planet continued. Between Piazzi's discovery date and 1808, stargazers found three more minor planets using Bode's Law: Pallas, Juno, and Vesta, each smaller than Ceres.

Astraea, the fifth asteroid, was discovered in 1845 and the fascination with asteroids as a new type of heavenly object began to build. In time, hundreds of asteroids would be found in the "asteroid belt" between Mars and Jupiter. Since those early days, new asteroids have been discovered almost every year.

Why did Doyle decide upon asteroid dynamics as Moriarty's most impressive work? Well, the asteroid discoveries triggered huge excitement in astronomy and the wider world. Swiftly enough there developed theories as to why asteroids were found between Mars and Jupiter and not a missing planet. By the time Doyle was writing *The Valley of Fear*, the hubbub about asteroids had died down as they were becoming commonplace. However, the first "near-earth" asteroid, Eros, was found in 1898. Today, in general, near-Earth objects (NEO's) such as Eros are small Solar System bodies whose orbit brings them into proximity with our planet. If the orbit of an NEO crosses that of the Earth, and the object is larger than 460 feet across, then the NEO is thought to be a potentially hazardous object (PHO). Please don't blame me for all these TLA's (three-letter acronyms). It's the way astronomers like to work, it seems. Anyhow, most known PHOs and NEOs are asteroids, with a tiny fraction being comets. (To be precise, there are over 29,000 known near-Earth asteroids (NEAs) and over a hundred known near-Earth comets (NECs); I did warn you about these TLA's.)

Maybe the situation was this: The discovery of Eros brings the question of asteroid dynamics much closer to home. It's certainly a dramatic development, even if astronomical events never play significant roles in Sherlock's cases. Perhaps the most intriguing thing about the astronomy choices in Doyle's tales is how they got there, but Doyle was certainly using such topics as a way of underlining the scientific literacy of Sherlock and others by making them knowledgeable about the current state of science.

Sherlock and Cryptology

In *The Adventure of the Dancing Men*, the tale begins when Sherlock is presented with a piece of paper upon which is set the following mysterious sequence of stick characters:

Though the solution of the cipher is the key element to cracking the case, Sherlock's first attempts at a solution are hampered by the fact that he has only the fifteen characters of the first message. Nonetheless, Sherlock assumes that this is a substitution code, with the most common character in the cipher representing the most common letter, E. "As you are aware," Sherlock says, "E is the most common letter in the English alphabet and it predominates to so marked an extent that even in a short sentence one would expect to find it most often.") He also notes that some of the characters had flags while some did not. Given the distribution of the characters that held flags and the fact that there is no spacing between the figures, or other obvious punctuation, Sherlock felt that the flags may be used to indicate the gap between words. He concludes that the first message (from Chicago gangster Abe Slaney to his former fiancé Elsie Cubitt) is "Am here Abe Slaney." Progress.

When four more messages are forthcoming, Sherlock once more assigns the most frequent of sixty-two characters to the letter E. He finds that there are seventeen Es among the sixty-two characters in the five messages. Next, he notes that the fourth message is just a five-letter word with E in both the second and fourth positions:

Sherlock notes that there's only a handful of words in the English language that could fit that pattern. "Sever," perhaps, or "lever." But Sherlock decides that "never" is the most likely solution, as he believes this is a conversation between two people, so a single word response of "never" seems the most likely. (Sherlockian scholars have been more imaginative, identifying thirty other solutions, including "seven," "renew," and "jewel.") Sherlock has now broken the cipher for the letters V and R. Further progress.

Next, Sherlock wonders whether Elsie's name might be hidden in a message. With this in mind, he spots two occurrences in the messages where there's a five-character pattern of E something something something E. As this clearly could be the name "Elsie," Sherlock takes a guess and assumes it to be the case, which means he now thinks he knows the characters for L, S, and I.

With the name Elsie logged, Sherlock could see that, in the second message, there's a single four-letter word that comes before Elsie's name and ending in the letter E. He assumes this is some kind of instruction or directions that the person writing the message is telling Elsie to do. As "come" seems the most fitting instruction, Sherlock now has the characters for the letters C and M.

Soon enough, Sherlock has cracked the whole code. The messages read:

1. AM HERE ABE SLANEY
2. AT ELRIGES
3. COME ELSIE
4. NEVER
5. ELSIE PREPARE TO MEET THY GOD

With some urgency, Sherlock notes that the fifth message, "Elsie, prepare to meet thy god," sounds more than a tad ominous. After rushing to the Cubitt home, Sherlock is too late. Hilton Cubitt is dead, slain by Slaney, and Elsie Cubitt, in despair over her husband's death, shoots herself in a vain attempt at suicide. Given this outcome, Sherlock's cipher solution can hardly be considered a triumph, but it's not too late for Sherlock to strike back. He lures Slaney to him by sending out a message in Slaney's own code, "Come here at once." Slaney, assuming the message is from his beloved Elsie, falls for the trap and is caught.

A little cryptographic postscript on this case. Sherlockian scholars have looked in detail at the dancing men and considered the potential cryptologic orientations of their arms and legs. They got 784 discrete possibilities of meaning using a straightforward substitution code. Furthermore, by variously inverting the dancing men, this doubles the total to 1,568 characters. How do they explain this apparent lapse of Sherlock the mastermind? Simply by assuming that Watson had replaced the actual code with one that was just difficult enough to fox the reader, but not too difficult to explain. For completion's sake, here is a crib sheet devised by Sherlockian scholar Michael J. Sare.

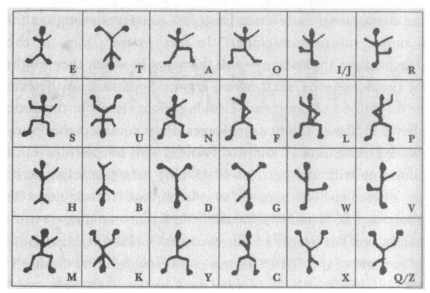

© Michael J. Sare

Sherlock and Mathematics

Sherlock delivers a math analogy for his deductive work from the get-go. In *A Study in Scarlet,* Watson is highly annoyed at a magazine article claiming that the conclusions of a trained observer are as "infallible as so many propositions of Euclid." Watson's reaction is intemperate. "What ineffable twaddle! . . . I never read such rubbish in my life . . . It is evidently the theory of some arm-chair lounger . . . I should like to see him clapped down in a third class carriage on the Underground, and asked to give the trades of all his fellow-travelers. I would lay a thousand to one against him." Sadly for Watson, it turns out that the author is Sherlock himself. "You would lose your money," Sherlock says. "As for the article, I wrote it myself."

Sherlock conjures up Euclid in *The Sign of Four* too. He lectures Watson on his writing style, laying into his companion for letting romanticism creep into his case report of *A Study in Scarlet.* Sherlock says, amusingly:

Honestly, I cannot congratulate you upon it. Detection is, or ought to be, an exact science, and should be treated in the same cold and unemotional manner. You have attempted to tinge it with romanticism, which produces much the same effect as if you worked a love-story or an elopement into the fifth proposition of Euclid.

Interesting here that Doyle assumes Watson's math knowledge is at least good enough to know that Euclid is a Greek mathematician living around 300 BC. He also that assumes his readers either know the same or are at least prepared to educate themselves in a public library (in those days before Google). Public libraries started to spring up in Britain after the Public Libraries Act in 1850. The Free Library Movement was one of the many groups in the Victorian era working for the "improvement" of the public through education.

There's more math in *The Sign of Four,* and Doyle once again assumes that his readers will research the background knowledge, supposing they don't already know it. The math concerned is known as "the Rule of Three." This is a rule that allows someone to solve problems based on the laws of proportions.

Here's how it works. Let's say we have three numbers, and let's call them a, b, and c. We use the letters of algebra, rather than the numbers themselves, as we need to generalize this math argument for any set of numbers. Let's also say that we have an unknown number, x (the unknown number is almost always "x" in algebra!), such that the algebraic relationship between our four numbers (a, b, c, and x) is given by:

$$a \ / \ b = c \ / \ x$$

In other words, the answer you get by dividing a by b is the same answer you get by dividing c by x. This is useful, as we can now calculate our unknown number, x, using the following rearranged relationship:

$$x = bc \, / \, a$$

In fact, we can rearrange the algebraic relationship to find *any* of the numbers at any time, assuming you know the value of the other three numbers, of course. That's the beauty of algebra. You may not have heard of the Rule of Three. That's because, in today's more mathematical world, it's thought to be so mathematically trivial that the term has gone out of use. While the Rule of Three was considered sufficiently important to warrant a name in Victorian times, Sherlock doesn't use it to make an actual calculation in *The Sign of Four*.

For those readers uncomfortable with the sight of algebra on these pages, it could be a lot worse, believe me. The origins of algebra, along with the word itself, come from a book written in the ninth century AD called *al-Kitāb al-Mukhtaṣar fī Ḥisāb al-Jabr wal-Muqābalah* (*The Compendious Book on Calculation by Completion and Balancing*) by Persian polymath Muḥammad ibn Mūsā al-Khwārizmī. If you carefully examine the Arabic title of al-Khwārizmī's great work, you will see the word al-jabr, meaning restoration or completion. Of course, the Anglicized version of the word al-jabr is algebra.

This book, which came to be known by the shortened title of al-jabr, is associated with the very specific math operation of taking something from one side of an equation and putting it on the other side of an equation. Algebra took on a lot of its modern ideas from this very important book, including the ones Sherlock uses, but the infamous use of algebraic symbols and letters implied by Sherlock and dreaded by school children the world over didn't begin with al-Khwārizmī's book. No, for the first few centuries until medieval times, algebra was practiced in prose. In short, longhand! Imagine how much more dreadful *that* would be in school. So, rather than writing a / b = c / x, Sherlock would have written something like, "imagine, Watson, that we have a quantity which when divided by a second quantity establishes a ratio which is exactly equal to a third quantity divided by a fourth."

Symbolic algebra, the version you know and "love" and which uses symbols such as +, -, %, <, >, and so on, didn't develop until centuries later. It's a surprise it took mathematicians that long, to be quite frank. Every time they carried out an operation, such as *adding*, they would write something like, "add number 7 to number 4." Or, for *dividing*, "divide the quantity 333 by the number 11," and so on. You can easily see that, if mathematicians were doing a long list of math operations, the time they spent writing out the operations would take longer than the time spent working out a solution! Thus, the symbols were adopted to avoid redundancy and save time.

Consider the equals sign, =. This was invented by Welsh physician and mathematician Robert Recorde in his 1557 book, *The Whetstone of Witte*. Recorde, who lived in Tenby, just an hour or so from my own home here in the south of Wales, explained (in medieval English) why using = was better than repeating the words "is equal to" over and over again:

> Howbeit, for easie alteration of *equations*. I will propounde a fewe exanples, bicause the extraction of their rootes, maie the more aptly bee wroughte. And to avoide the tediouse repetition of these woordes : is equalle to : I will sette as I doe often in woorke use, a pair of paralleles, or Gemowe lines of one lengthe, thus: =====, bicause noe .2. thynges, can be moare equalle.

Sherlock might not have used the Rule of Three to make a calculation in *The Sign of Four*, but he *did* use the rule in *The Adventure of the Musgrave Ritual.* To recap this adventure, Sherlock was able to follow the directions of the Musgrave Ritual to locate a small cellar room in which the ancient crown of the King of England had allegedly been hidden. Sherlock's method of detection in this case was using geometry and the Rule of Three.

The directions were given as a step count, "north by ten and by ten, east by five and by five, south by two and by two, west by one and by one, and so under." But the main problem of the puzzle was where to begin. Sherlock had worked out that the starting point of the directions was the tip of the shadow of the elm tree at a particular position of the Sun, yet the said elm tree was no more, felled by lightning a decade before.

Sherlock uses the fact that Reginald Musgrave knew that the elm was sixty-four feet high. Then, to work out the length of the elm's shadow from its height, Sherlock puts up a six-foot fishing pole at the site of the elm stump. He then finds that the shadow of said fishing pole, when the Sun was in the necessary position, was nine feet long. This now allows Sherlock to set up the proportions of the Rule of Three and calculate the elm's shadow-length to be ninety-six feet.

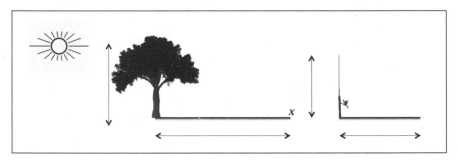

Having calculated the elm's shadow-length, Sherlock knows he needs to start at a distance of ninety-six feet from the stump in the same direction as the shadow of the pole. Sadly, when he finds the small room where the King of England's crown is meant to be concealed, Sherlock is foiled. For, instead of the crown, he finds the body of Brunton the butler who had followed the ritual before Sherlock.

There are botanical objections to the above math. The algebra for working out the solution depends on the elm tree height still being the same in Sherlock's day as it was two hundred and fifty years ago when the crown was concealed. Needless to say, elm trees can grow higher than sixty-four feet. (For example, the largest elm tree in the US, known as the Rathbone Elm, was taken down in 1959 because of Dutch elm disease. It was believed to be four hundred years old, with a height of ninety-nine feet and spread of 150 feet.)

There's climate science to consider too. Soil and climate are crucial ingredients in a tree's adult height, and this is an old tree, at least two hundred and fifty years old. How would it have fared against damaging storm winds and lightning strikes? One last factor to figure into the calculation is Sherlock's own height. Average human height was shorter when the crown was hidden all those years ago. Did Sherlock factor into his algebra some kind of adjustment for this? Did he adjust his stride? The best kind of calculation (that is, a successful one!) would have done so.

In fact, the question of calculating human height from stride length is a math that Sherlock performs in *A Study in Scarlet*. Sherlock says of the math that "it's a simple calculation enough," but let's look at just how simple it is to find a suspect's height from his stride length. In *A Study in Scarlet*, Sherlock analyzes the scene of the crime where Enoch Drebber's body was discovered. Next, Sherlock provides Inspectors Gregson and Lestrade with several clues. Chief among these is that the perpetrator is taller than six feet. The inspectors are skeptical, as is Watson, though admittedly one of the reasons for their skepticism and incredulity is that Doyle treats us to one of Sherlock's full onslaughts of deduction:

There has been murder done, and the murderer was a man. He was more than six feet high, was in the prime of life, had small feet for his height, wore coarse, square-toed boots and smoked a Trichinopoly cigar. He came here with his victim in a four-wheeled cab, which was drawn by a horse with three old shoes and one new one on his off fore leg. In all probability the murderer had a florid face, and the finger-nails of his right hand were remarkably long. These are only a few indications, but they may assist you.

Later, Watson queries Sherlock's method of deducing the killer's height. "You amaze me, Holmes. Surely you are not as sure as you pretend to be of all those particulars which you gave." Sherlock's reply includes the detail: "Why, the height of a man, in nine cases out of ten, can be told from the length of his stride. It is a simple calculation enough, though there is no use my boring you with figures." Yet good Sherlockians demand to be bored with the figures!

Many Sherlockians believe the stride/height calculation to be meaningless. The human stride varies with conditions and circumstance, they object. But such calculations as those done by Sherlock are still everywhere to be found. A quick Google search will provide you with a simple enough calculation, for example, that human height, male or female, is roughly 2.4 times the stride length. Elsewhere, the formula is slightly different. In Doyle's early tales, he has Sherlock use some kind of magical formula for the stride/height calculation. Sherlock once more claims to be able to work out Jonathon Small's height from the length of his stride in *The Sign of Four*. Similarly, in *The Boscombe Valley Mystery*, Sherlock refers to the case's murderer as a "tall man" and tells Watson that the rough estimate is based on stride length.

By the time of Doyle's sixth tale, Sherlock has somewhat retracted his earlier bold claims about his stride/height formula. Finally, after using the math in three of the first half dozen tales, he never uses it again. And it's easy to see why. Contemporary forensic science places far more emphasis on foot size than stride length to estimate human height. Or, as the FBI puts it:

> Contrary to the plotting of detective fiction, it isn't possible to estimate someone's height by the distance between steps, his gait, because during the commission of a crime, a suspect is usually moving very fast; he is running or backing up or moving sideways or struggling, attacking or defending, even sneaking around. The thing he isn't doing is moving normally.

Sherlock and Ballistics

Unlike his outdated human height calculations, Sherlock's use of ballistics is still something used today in forensics. His first conspicuous use of ballistics occurs in *The Adventure of the Reigate Squire*. The basic relevant plot is as follows: A coachman is found dead. His name is William Kirwan, and he works for a well-to-do family known as the Cunninghams, whose estate is nearby the Acton estate in Surrey where Sherlock and Watson are resting after a rather strenuous case in France.

Both Cunningham senior and junior claim to have seen the murderer. Cunningham junior, Alec, is described by Watson as a "dashing young fellow, whose bright, smiling expression and showy dress were in strange contrast with the business which had brought us there." Young Cunningham is sarcastic and arrogant and treats Sherlock with smug condescension, an early indication that all might not be well, as one thing the English simply can't abide is bad manners.

According to Alec Cunningham, coachman Kirwan and his attacker were locked in combat when the fatal shot was fired. The murderer then fled. On a mere cursory examination of the body, Sherlock immediately concludes that Alec Cunningham is a liar.

The evidence? Simply this. There's no gunpowder mark on the corpse. When the case is almost at closure, Sherlock explains. The obvious lack of any powder blackening had persuaded him that the fatal shot had been fired from a greater distance than four yards. Based on those ballistics, Sherlock worked a case against the actual murderers, the arrogant Cunninghams.

There's another conspicuous use of ballistics in *The Adventure of the Dancing Men*. This is the tale we featured on page 177 about solving the cipher of the messages in the strange form of dancing men figures. At first, it seems that Elsie Cubitt shot and murdered Milton, her husband, then failed in her attempted suicide. Needless to say, Sherlock the science virtuoso didn't buy it. The fated husband got in touch with Sherlock about the strange cipher. Sherlock cracked the code and knew there was a third party in this intriguing case. What's more, the two servants who had found the Cubitts, one dead and the other on her way, said that they could instantly smell gunpowder after hearing shots and charging downstairs to the scene of the crime in the study.

Sherlock's conclusion? That both study door and study window had been open. The key science here is the process of diffusion, the action of a substance spreading through another and in many directions. In this particular case, of course, the spread is the movement of gunpowder vapor through the air of the house. We're all aware of diffusion. The delicious smell of strawberries from the kitchen when you are lying in bed, say, or the browning of hot water when a tea bag is dipped in a cup, or the tiny particles of dust or smoke and the way they diffuse into the air and cause pollution.

Doyle's countrymen had been key to the study of diffusion in physics during the nineteenth century. Scottish chemist Thomas Graham pointed out in 1833, "gases of different nature, when brought into contact, do not arrange themselves according to their density, the heaviest undermost, and the lighter uppermost, but they spontaneously diffuse, mutually and equally, through each other, and so remain in the intimate state of mixture for any length of time." Graham's work led Scottish mathematician James Clerk Maxwell in 1867 to calculate the coefficient of diffusion for CO_2 in the air. The error rate was less than 5 percent.

As it was originally formulated, Graham's law of diffusion attempts to model the way in which two gases mix. This is the case in *The Adventure of the Dancing Men*. One of the gases is gunpowder vapor and the other is the general air of the house. The powder vapor must not only meet with the air, but also move *through* the air and arrive upstairs almost instantly. Sherlock is aware that this process won't occur without a breeze. Wind-assisted, like some times in the one-hundred-meter dash at the Olympic games. It's a typically astute observation by Sherlock, but Inspector Martin is not buying it at first:

> **Sherlock:** "You remember, Inspector Martin, when the servants said that on leaving their room they were at once conscious of a smell of powder I remarked that the point was an extremely important one?"
>
> **Martin:** "Yes, sir; but I confess I did not quite follow you."
>
> **Sherlock:** "It suggested that at the time of the firing the window as well as the door of the room had been open. Otherwise the fumes of powder could not have been blown so rapidly through the house. A draught in the room was necessary for that. Both door and window were only open for a very short time, however."

Feeling sure that the window had been open at the time of the tragedy, Sherlock wondered if there might have been a third person in the affair. Someone who stood outside and fired in. Any shot directed at this third person might miss. Sherlock looked, and there, lo and behold, a bullet mark was found on the windowsill. Sherlock then proceeds to describe the murder scene in more detail. The third person outside the window was Abe Slaney. He and Hilton Cubitt fired almost simultaneously, making the very loud noise which woke the staff upstairs. Slaney killed Cubitt, but his bullet missed Slaney and got stuck in the sill. Elsie Cubitt, devastated by her husband's death, then shot herself in the head. A legendarily bad shot.

Sherlock and the Underground

There had always been rumors of a world under London's ground. Subterranean chambers and tunnels had been reported. One linked the crypt of St. Bartholomew the Great with Canonbury. Another ran the shorter distance between the nunnery and priory of Clerkenwell. Then, there were the sweeping catacombs of Camden Town, hidden beneath the Camden goods yard. Roman temples had been found within this latent London. Statues of ancient deities discovered, and in a condition which suggested, mysteriously, that they were deliberately buried that way.

Investigating his cases, Sherlock climbed down more ladders to explore the old, buried town than he had toiled up City staircases. One of the characteristic drawings of Sherlock's city was that of its horizontal levels. The schema showed detail in depth, from the rooftops of its grand houses down to the caverns of its sewers, and the whole edifice bore down upon and almost crushed each level with its weight. It was commonly said that, in the city's history, none who knew London well would deny that its treasures must be sought in its depths. It was an ambiguous thought, one which equally applied to social as well as topographical circumstance. Sherlock's Baker Street Irregulars were society's "underground."

The railway mania of the 1840s found its echo underground too. The invention of the steam engine and the specter of indus-trialization led London's burgeoning development of a commuting population to traffic congestion problems. Horse manure became a major concern. Sherlock's London typically had over ten thousand cabs and several thousand omnibuses. Each mode of transport used several horses, so his city boasted more than fifty thousand horses in public transport alone, with each animal making fifteen to thirty-five pounds of manure a day. One commentator made the rather Darwinian remark, "How much pleasanter the streets of a great city would be if the horse was an extinct animal."

The sweepers' job was to clear paths through the dung, which was usually sludge in the wet weather of London, or a fine blown powder on the odd dry day. Alas, the piles of manure attracted vast numbers of flies. One estimate suggests that three billion flies hatched in horse manure per day in such cities, with tens of thousands of deaths each year blamed on the manure.

Some had worse jobs. The horses produced tens of thousands of gallons of urine daily, which had to be dealt with. The creatures were amazingly noisy, their iron shoes on cobbles made conversation intolerable on the bustling streets and were far more dangerous than modern motorized traffic, with a fatality rate three-quarters higher per capita than today. Worse still, the problems with these creatures didn't disappear when they died. The average working horse had a life expectancy of only three years. Scores died each day and, as dead horses were hard to shift, street cleaners would wait days for the corpses to rot so that they could more easily be sawed into chunks.

In short, Sherlock's London had been a city desperately looking for a traffic solution, and they found it in the train. The train was hailed as an environmental savior. By the middle of the century, there were seven railway termini located around the urban center of the metropolis, and soon the idea arose of an underground railway to link the City of London with these satellite stations. We are now well familiar with London's long-established Underground. The "tube" was the world's first subterranean railway, and today over one hundred miles of underground and networked track serve around four million passengers daily, one of the largest on the planet.

Conan Doyle took advantage of this subterranean zeitgeist. Published in 1908, *The Adventure of the Bruce-Partington Plans* is one of those Doyle tales where Sherlock's casework also features Mycroft Holmes. The plot has it that plans to a new submarine have gone missing. A body is found next to the London Underground train tracks near Aldgate station. The body is that of Arthur Cadogan West, a junior clerk at the Royal Arsenal in Woolwich, and is found along with stolen top-secret submarine schematics. At the urging of Mycroft, Sherlock investigates the circumstances of Cadogan West's death.

Doyle dials up the drama in one of his most spectacular stories of espionage. Sherlock is taxed by the fact that no ticket for the train is found on the dead body:

> **Sherlock:** "Why had he no ticket?"
> **Inspector Lestrade:** "The ticket would have shown which station was nearest the agent's house. Therefore he took it from the murdered man's pocket."
> **Sherlock:** "Good, Lestrade, very good. Your theory holds together. But if this is true, then the case is at an end. On the one hand, the traitor is dead. On the other, the plans of the Bruce Partington submarine are presumably already on the Continent. What is there for us to do?"
> **Mycroft:** "To act, Sherlock—to act! All my instincts are against this explanation. Use your powers! Go to the scene of the crime! See the people concerned! Leave no stone unturned! In all your career you have never had so great a chance of serving your country."

Sherlock's intuition about the missing train ticket helps him deduce that Cadogan West was never actually *in* the train, but *on top* of it. Furthermore, Cadogan West was murdered in an apartment that sat near the train tracks around Gloucester Road Station, one of the few areas where the tube train surfaces. The killer then simply dumped the body onto the train's top when it was briefly stationary. What's more, the murderer may have had the idea after hearing train station guards along the Underground shouting out that it is forbidden to ride on the roof. There, the body of Cadogan West remained for approximately twelve stops—until physics had its way with him. A combination of friction, inertia, and momentum, to be precise.

First, the friction. The dead weight of the body in contact with the roof surface of the train would resist any movement. As Newton said in his first law of motion, a body tends to remain at rest, or in uniform motion, unless it is compelled to move. It has inertia, a little like a teenager playing on a PlayStation when asked to help with the housework. The impetus that the dead weight of Cadogan West needs is provided by two factors near Aldgate in what Sherlock refers to as "points, and a curve," for Aldgate is a junction, and the train tracks are curved. When the train runs over the "points" at a junction like Aldgate, the ride becomes bumpier, and the contact between the train roof and Cadogan West is lessened a little. The corpse then slides. This reduction of friction, along with the body's momentum in a straight line as the train goes around a curve, meant that Cadogan West's slide continued until his dead weight was trackside. Pure physics. Indeed, had the train continued smoothly in a never-ending straight line, Cadogan West would have forever remained on the roof (a little like Tom Cruise's hitman character Vincent in Michael Mann's marvelous 2004 movie, *Collateral*—a dead man riding unnoticed on a train).

Sherlock and the End

One case that has never been cracked is that of the fate of Sherlock himself. Almost 140 years after a young doctor made notes for a novel that became *A Study in Scarlet*, the world's greatest-ever consulting detective continues to elude any final verdict.

Sure, we know that Sherlock retired from Baker Street to the Sussex Downs to keep bees. Conan Doyle told us so. In Doyle's 1905 tale, *The Adventure of the Second Stain*, Watson lets slip that his companion had been pensioned off to a pastoral idyll. As we saw before, Conan Doyle once tried killing Sherlock off. Ever eager, it seems, to sweep Sherlock offstage, maybe retirement seemed a kinder outcome than the downright murder of his famous fictional detective.

A dozen years later and Doyle drops another hint. The tale, *His Last Bow: An Epilogue of Sherlock Holmes*, has been seen as a propaganda tool intended to boost morale for British readers. Sherlock and Watson are pitted against a German spy ring. Set in August 1914, on the eve of WWI, it brings to an end the Sherlockian chronology created by Conan Doyle.

His Last Bow is a rare treat for a number of reasons. It's told in the third person, and it's Doyle's only attempt to show his dynamic duo in action in their later years: "Stand with me here upon the terrace, for it may be the last quiet talk we shall ever have," Sherlock says to Watson. After *His Last Bow*, whenever Doyle did condescend to write the odd Sherlock tale, he opted for nostalgia pieces set in the detective's past. Maybe he felt that the young readers of the roaring and automotive 1920s weren't interested in reading about the old boys of Baker Street.

This lack of detail in Sherlock's ultimate fate is typical of Doyle. Little wonder the Sherlockian phenomenon persists. In his sixty tales, Doyle may have slapped on some limited detail, but he otherwise left vast swathes of the canvas blank. Here's another factor in the creation of the casual artistic brilliance of the Sherlock stories: They incorporate a buff but teasingly unfinished fictional universe. Think about Doyle's characters, potent but finished only as pencil portraits, perhaps. These characters coast in and out of Conan Doyle's tales in a mere matter of paragraphs. Mycroft, the even-smarter older brother, was mentioned only in four stories, yet he made an enormous impression on the literary world. Moriarty, the diabolical mathematician and criminal mastermind, appears directly only in two tales and mentioned tangentially in five others, yet is a fundamental influence on all secretive and superintelligent nemeses ever since. Irene Adler appears in just a single story, yet is the femme fatale who bested Sherlock.

Then, there's the unreported Sherlock cases referenced by Watson and which sound terribly more exciting than the one we're reading: *The Affair of the Vatican Cameos*; *The Politician, the Lighthouse, and the Trained Cormorant*; *A Full Account of the Club-Foot and his Abominable Wife*! The chronology of the Doyle canon is esoteric enough to make the impossible possible. Ever since Conan Doyle laid down his quill, others have tried to fill the void in his universe.

There are so many reasons why Sherlock lives on. Since his arrival in 1887, the fiftieth year of Queen Victoria's reign, Sherlock Holmes has been little short of a literary miracle. Conan Doyle died of a heart attack in 1930, three years after the last Sherlock tale appeared in *The Strand*. He was seventy-one years of age. A creative genius, maybe. A prolific inventor gifted with a huge appetite for life, perhaps. Yet a mere mortal who would ultimately lose the battle with his superhuman paper creation, who had never lived, and will never die.

THE END

CHRONOLOGY OF ARTHUR CONAN DOYLE

1859

Arthur Ignatius Conan Doyle born on May 22 at 11 Picardy Place, Edinburgh, to Charles Altamont Doyle and Mary Doyle (née Foley)

1860

Sidney Paget, Doyle's illustrator for *The Strand*'s Sherlock stories, born October 4, London, England

1868

Conan Doyle moves south and attends a Roman Catholic preparatory school, Hodder House, in Stonyhurst, Lancashire, England

1870–1875

Conan Doyle attends Stonyhurst College, a Catholic school run by Jesuits

1874

Jean Leckie, Conan Doyle's second wife, born on March 14 at 3 Kidbrook Terrace, Kidbrook, Kent

1876–1881

Conan Doyle studies medicine at the University of Edinburgh

1877

Conan Doyle meets Scottish surgeon and lecturer Dr. Joseph Bell, whom he identifies as the intellectual model for his detective Sherlock Holmes

1879

In September and October, Conan Doyle's first publications, *The Mystery of Sasassa Valley* and *Gelsenium as a Poison*, appear in *Chamber's Edinburgh Journal* and the *British Medical Journal*

1880

From February to September, Conan Doyle is employed as ship's surgeon on the Greenland whaler *Hope*

1881

Conan Doyle graduates from the University of Edinburgh with a bachelor of medicine and master of surgery; makes his second voyage as ship's surgeon on *S. S. Mayumba*, sailing to Africa's west coast

1882

In June, Conan Doyle establishes medical practice at 1 Bush Villas, Elm Grove, in South-sea, Portsmouth; writes his first novel, *The Narrative of John Smith* (lost but posthumously published in 2011)

1885

Conan Doyle awarded MD by the University of Edinburgh (dissertation title: *On Vasomotor Influences in Tabes Dorsalis*); on August 5, marries Louisa Hawkins ("Touie") in St. Oswald's Church, Thornton, North Yorkshire

1886

Conan Doyle writes the novella *A Study in Scarlet*, introducing Sherlock Holmes. The book is declined by three publishers before Ward, Lock & Co. buy the copyright for £25 and agree to publish the following year

1887

A *Study in Scarlet* appears in Beeton's Christmas Annual

1888

First edition of *A Study in Scarlet* with six illustrations by Charles Doyle is published by Ward, Lock & Co; between August 31 and November 9, the five canonical victims of Jack the Ripper are murdered in the East End of London

1889

Conan Doyle and Oscar Wilde are commissioned by Joseph Stoddart to write for *Lippincott's Monthly Magazine*; Wilde's novel is *The Picture of Dorian Gray*

1890

In February, *Lippincott's Monthly Magazine* publishes *The Sign of Four*; Conan Doyle studies ophthalmology in Vienna and travels Europe; in December, George Newnes launches *The Strand Magazine*

1891

Conan Doyle writes the anonymous *The Voice of Science* story for *The Strand*, and the first set of Sherlock Adventures (July 1891–June 1892); Doyle moves to London to practice as an eye specialist at Upper Wimpole Street but by August gives up medicine to make a living as an author

1892

The first set of Adventures is published in book form as *The Adventures of Sherlock Holmes*; a second set of Adventures is published in *The Strand* (December 1892–December 1893)

1893

Conan Doyle visits the Reichenbach Falls; Doyle's wife Louisa is diagnosed with tuberculosis; *The Final Problem*, in which Sherlock "dies," appears in December, beginning the period known as the "Great Hiatus" (in Sherlockian chronology, Sherlock's death is in May 1891 and his "resurrection" in *The Empty House* is in April 1894); the first adaptation of Sherlock for the theater, *Under the Clock*, opens at Royal Court Theatre, London, in November

1894

The Memoirs of Sherlock Holmes, the second set of Adventures, publishes in book form

1897

Conan Doyle moves with his wife Louisa to Undershaw, Surrey; meets Jean Leckie

1899

In November, American actor William Gillette plays Sherlock on stage in *Sherlock Holmes, or the Strange Case of Miss Faulkner*, at Garrick Theater, New York

1900

Conan Doyle serves as a volunteer during the Boer War at the Langman Field Hospital; stands unsuccessfully for Parliament as a Liberal Unionist in the Edinburgh Central constituency

1901

Sherlock reemerges in *The Hound of the Baskervilles*, a retroadventure serialized in *The Strand* (August 1901–April 1902)

1902

Conan Doyle is knighted by King Edward VII for penning a pamphlet which justified British actions during the Boer War

1903

A new series of thirteen stories appears in Collier's Weekly (September 1903–January 1905) and *The Strand* (October 1903–December 1904); *The Adventure of the Empty House* explains what happened at the end of *The Final Problem*

1905

Latest series of Sherlock tales is published in book form as *The Return of Sherlock Holmes*

1906

July 4, Louisa Doyle dies of tuberculosis

1907

On September 18, Conan Doyle marries Jean Leckie at St. George Hanover Square, Middlesex, England

1910

In June, a theatrical adaptation of *The Speckled Band* is written and produced by Conan Doyle and performed at the Adelphi Theatre, London

1914

The Strand serializes *The Valley of Fear* (September 1914–May 1915)

1915

The parody series *Herlock Sholmes*, written by Frank Richards, first appears and runs until 1954 in boys' magazines including *The Greyfriars Herald*, *The Magnet*, and *The Gem*; *The Valley of Fear* publishes in book form

1917

Publication of *His Last Bow*, including preface and seven stories in *The Strand* between 1908 and 1913; publication of *The Adventure of the Cardboard Box*

1921

From now until 1923, *Stoll Picture Productions* produce forty-seven film adaptations, two being feature-length

1924

Publication of *Memories and Adventures*

1927

The Case-Book of Sherlock Holmes publishes, comprised of a dozen tales published in *The Strand* and various US magazines between 1921 and 1927; *The Adventure of Shoscombe Old Place* is the last Sherlock tale published by Conan Doyle

1928

John Murray publishes *Sherlock Holmes: The Complete Short Stories*

1929

John Murray publishes *Sherlock Holmes: The Complete Long Stories*; Basil Dean directs *The Return of Sherlock Holmes*, the first Sherlock film to feature sound

1930

July 7, Conan Doyle dies at Windlesham Manor in Crowborough, East Sussex, England

1934

New York's *Baker Street Irregulars* is set up by Christopher Morley; London's *Sherlock Holmes Society* is founded by Dorothy Sayers et al.

1939–1946
English actor Basil Rathbone plays Sherlock in fourteen films (and many radio productions), remixing Sherlock to a more modern setting; first film is *The Hound of the Baskervilles* and last is *Dressed to Kill*

1946
The Baker Street Irregulars launch the *Baker Street Journal*

1951
Sherlock's Baker Street fictional residence is created for display at the Festival of Britain, and later transferred to the Sherlock Holmes pub, Northumberland Street, London

1954
Adrian Conan Doyle and John Dickson Carr publish *The Exploits of Sherlock Holmes*

1964–1968
BBC broadcast television series *Sherlock Holmes* casts Douglas Wilmer (1964–1965) and Peter Cushing (1968) as Sherlock and Nigel Stock as Watson

1970
Billy Wilder directs *The Private Life of Sherlock Holmes*, with Robert Stephens as Sherlock and Colin Blakely as Watson

1976
Herbert Ross directs Oscar-nominated *The Seven Per-Cent Solution* with Nicol Williamson as Sherlock, Robert Duvall as Watson, and Laurence Olivier as Moriarty

1978

Michael Dibdin's Sherlock pastiche novel *The Last Sherlock Holmes Story* publishes, in which Sherlock attempts to solve the Jack the Ripper murders, suspecting the culprit to be his nemesis, James Moriarty

1979

Bob Clark directs *Murder by Decree* with Christopher Plummer as Sherlock and James Mason as Watson, who are investigating the Ripper murders. The film's plot is influenced by the book *Jack the Ripper: The Final Solution* (1976) by Stephen Knight, who presumed that the killings were part of a Masonic plot

1984–1994

Granada Television series *Sherlock Holmes* (with nine series varying in title according to the source material) stars Jeremy Brett as Sherlock and David Burke and Edward Hardwicke as Watson. Of the sixty Sherlock tales written by Conan Doyle, a full forty-three were adapted by Granada in the series, spanning thirty-six one-hour episodes and five feature-length specials. Brett's portrayal remains very popular and is considered by many to be the definitive on-screen version of Sherlock. Watson is played as a competent associate, as in Doyle's tales

1988

Thom Eberhardt directs *Without a Clue* with Michael Caine as Sherlock and Ben Kingsley as Watson, in which Watson is the brilliant detective and "Sherlock Holmes" is a hired actor posing as the detective so that Watson can protect his reputation as a doctor

1990

Opening of the Sherlock Holmes Museum on Baker Street

1999–2001
Animated television series, *Sherlock Holmes in the 22nd Century*, in which Sherlock is brought back to life in the 22nd century

2000
The last of Conan Doyle's Sherlock tales enter the public domain

2002–
Ukrainian video-game development studio, *Frogwares*, launch their series of adventure games based on Conan Doyle's Sherlock tales. While the franchise is based on Doyle's stories, each game has an original plot and storyline, including the 2009 release, *Sherlock Holmes Versus Jack the Ripper*, and the 2021 release, *Sherlock Holmes Chapter One*, in which a young Sherlock investigates a mystery in his family's home on the Mediterranean island of Cordona after his mother's death

2005
Mitch Cullin's *A Slight Trick of the Mind* is published in which a ninety-three-year-old beekeeping Sherlock struggles to recall the details of his final case because his mind is slowly deteriorating; the book is adapted for film as *Mr. Holmes* (2015) in which Sherlock is played by Sir Ian McKellen

2009
Guy Ritchie directs *Sherlock Holmes* with Robert Downey Jr. as Sherlock and Jude Law as Watson, in which the pair are hired by a secret society to foil a mysticist's plot to gain control of Britain by seemingly supernatural means

2010
Opening of the Arthur Conan Doyle Collection (Lancelyn Green Bequest), Portsmouth City Library

2010–2017

The BBC series *Sherlock* is written and produced by Mark Gatiss and Steven Moffat, featuring Benedict Cumberbatch as Sherlock and Martin Freeman as Watson. Thirteen episodes have been produced, with four three-part series broadcast between 2010 to 2017, and a one-off special episode that broadcast on January 1, 2016. *Sherlock* is set in the present day, while the special episode features a Victorian period fantasy resembling the original Doyle stories

2011

Anthony Horowitz's *The House of Silk* is officially endorsed by the Arthur Conan Doyle estate as an authorized pastiche; Guy Ritchie directs *Sherlock Holmes: A Game of Shadows*. The film follows an original premise but also incorporates elements of Conan Doyle's tales, including *The Final Problem* and *The Adventure of the Empty House*

2012–2019

CBS series *Elementary* stars Jonny Lee Miller as Sherlock and Lucy Liu as a female Watson

2014–2015

Sherlock Holmes: The Man Who Never Lived and Will Never Die becomes a major exhibition at the Museum of London

2015

A "lost" Sherlock story, *Sherlock Holmes: Discovering the Border Burghs and, by deduction, the Brig Bazaar*, is found in Selkirk, on the Scottish Borders, and reprinted in the *Daily Telegraph*; the tale's attribution to Conan Doyle is largely dismissed. Doyle wrote the story in 1904 to raise money for a bridge in Selkirk

2018

Etan Cohen directs *Holmes and Watson*, a "comedy" starring Will Ferrell as Sherlock and John C. Reilly as Watson

CHRONOLOGY OF SHERLOCK HOLMES STORIES

#	Story	Published	Date Story Set
1.	*A Study in Scarlet*	Nov 1887	Mar 1881
2.	*The Sign of Four*	Feb 1890	Jul/Sep 1888

The Adventures of Sherlock Holmes

#	Story	Published	Date Story Set
3.	*A Scandal in Bohemia*	Jul 1891	Mar 1888
4.	*The Red-Headed League*	Aug 1891	Oct 1890
5.	*A Case of Identity*	Sep 1891	During 1888
6.	*The Boscombe Valley Mystery*	Oct 1891	Jun 1888
7.	*The Five Orange Pips*	Nov 1891	Sep 1887
8.	*The Man with the Twisted Lip*	Dec 1891	Jun 1889
9.	*The Adventure of the Blue Carbuncle*	Jan 1892	Dec 1889
10.	*The Adventure of the Speckled Band*	Feb 1892	Apr 1883
11.	*The Adventure of the Engineer's Thumb*	Mar 1892	Summer 1889
12.	*The Adventure of the Noble Bachelor*	Apr 1892	During 1887
13.	*The Adventure of the Beryl Coronet*	May 1892	Feb 1890
14.	*The Adventure of the Copper Beeches*	Jun 1892	During 1890

The Memoirs of Sherlock Holmes

15.	The Adventure of Silver Blaze	Dec 1892	Sept, 1890
16.	The Adventure of the Cardboard Box	Jan 1893	During 1888
17.	The Adventure of the Yellow Face	Feb 1893	During 1888
18.	The Adventure of the Stockbroker's Clerk	Mar 1893	June 1888
19.	The Adventure of the Gloria Scott	Apr 1893	During 1875
20.	The Adventure of the Musgrave Ritual	May 1893	During 1879
21.	The Adventure of the Reigate Squire	Jun 1893	April, 1887
22.	The Adventure of the Crooked Man	Jul 1893	Summer 1888
23.	The Adventure of the Resident Patient	Aug 1893	Oct 1881
24.	The Adventure of the Greek Interpreter	Sep 1893	During 1888
25.	The Adventure of the Naval Treaty	Oct 1893	July 1888
26.	The Adventure of the Final Problem	Dec 1893	During 1891
27.	The Hound of the Baskervilles	Aug 1901	During 1889

The Return of Sherlock Holmes

28.	The Adventure of the Empty House	Sept 1903	April 1894
29.	The Adventure of the Norwood Builder	Oct 1903	During 1894
30.	The Adventure of the Dancing Men	Nov 1903	During 1898
31.	The Adventure of the Solitary Cyclist	Dec 1903	During 1895
32.	The Adventure of the Priory School	Jan 1904	May 1901
33.	The Adventure of Black Peter	Feb 1904	During 1895
34.	The Adventure of Charles Augustus Milverton	Apr 1904	During 1899
35.	The Adventure of the Six Napoleons	May 1904	During 1900
36.	The Adventure of the Three Students	Jun 1904	During 1895
37.	The Adventure of the Golden Pince-Nez	Jul 1904	During 1894
38.	The Adventure of the Missing Three-Quarter	Aug 1904	During 1896
39.	The Adventure of the Abbey Grange	Sep 1904	During 1897
40.	The Adventure of the Second Stain	Dec 1904	July 1888

His Last Bow

41.	*The Adventure of Wisteria Lodge*	Sep 1908	During 1892
42.	*The Adventure of the Bruce-Partington Plans*	Dec 1908	Nov 1895
43.	*The Adventure of the Devil's Foot*	Dec 1910	Spring 1897
44.	*The Adventure of the Red Circle*	Mar 1911	1895–1902
45.	*The Disappearance of Lady Frances Carfax*	Dec 1911	Unknown
46.	*The Adventure of the Dying Detective*	Dec 1913	Feb 1890
47.	*His Last Bow*	Sep 1917	Aug 1914
48.	*The Valley of Fear*	Sep 1914	Jan 1889

The Case-Book of Sherlock Holmes

49.	*The Adventure of the Mazarin Stone*	Oct 1921	During 1903
50.	*The Problem of Thor Bridge*	Feb 1922	During 1900
51.	*The Adventure of the Creeping Man*	Mar 1923	During 1903
52.	*The Adventure of the Sussex Vampire*	Jan 1924	During 1896
53.	*The Adventure of the Three Garridebs*	Jan 1925	During 1902
54.	*The Adventure of the Illustrious Client*	Feb 1925	During 1902
55.	*The Adventure of the Three Gables*	Oct 1926	During 1903
56.	*The Adventure of the Blanched Soldier*	Nov 1926	Jan 1903
57.	*The Adventure of the Lion's Mane*	Dec 1926	During 1907
58.	*The Adventure of the Retired Colourman*	Jan 1927	During 1899
59.	*The Adventure of the Veiled Lodger*	Feb 1927	During 1896
60.	*The Adventure of Shoscombe Old Place*	Apr 1927	During 1902

(More Detailed) Chronology of Sherlock Holmes Stories

1875

The Adventure of the Gloria Scott is framed within a story set some time after the events of *A Study in Scarlet*

1879

The Adventure of the Musgrave Ritual

1881

Friday March 4, *A Study in Scarlet*

1883

Early April, *The Adventure of the Speckled Band*

1887

Mid-April, *The Adventure of the Reigate Squire*
Late September, *The Five Orange Pips*

1888

Saturday January 7, *The Valley of Fear*
Tuesday March 20, *A Scandal in Bohemia*
July or September, *The Sign of Four*
October, *The Adventure of the Noble Bachelor*

1889

Summer, *The Adventure of the Engineer's Thumb*
Summer, *The Adventure of the Crooked Man*
Early June, *The Boscombe Valley Mystery*

Wednesday June 19, *The Man with the Twisted Lip*
July, *The Adventure of the Second Stain*
July, *The Adventure of the Naval Treaty*
Early October, *The Hound of the Baskervilles*

1890
Thursday October 11, *The Adventure of the Red-Headed League*

1891
Friday April 24, *The Adventure of Wisteria Lodge*
1892
Late March, *The Adventure of Wisteria Lodge*

1894
Early April, *The Adventure of the Empty House*
August, *The Adventure of the Norwood Builder*
Late November, *The Adventure of the Golden Pince-Nez*

1895
Saturday April 23, *The Adventure of the Solitary Cyclist*
Mid-year, *The Adventure of the Three Students*
Early July, *The Adventure of Black Peter*
Thursday November 21, *The Adventure of the Bruce-Partington Plans*

1896
Late year, *The Adventure of the Veiled Lodger*

1897
January, *The Adventure of the Abbey Grange*
Tuesday March 16, *The Adventure of the Devil's Foot*

1898
Summer, *The Adventure of the Retired Colourman*
July, *The Adventure of the Dancing Men*

1901

Thursday May 16, *The Adventure of the Priory School*

1902

Late June, *The Adventure of the Three Garridebs*
Wednesday September 3, *The Adventure of the Illustrious Client*

1903

January, *The Adventure of the Blanched Soldier*
Sunday September 6, *The Adventure of the Creeping Man*

1907

Late July, *The Adventure of the Lion's Mane*

1914

Sunday August 2, *His Last Bow*

INDEX

Note: Characters are alphabetized by first name.